EXAM *Revision* NOTES

AS/A-LEVEL
Geography

Michael Raw

2nd Edition

Philip Allan Updates, an imprint of Hodder Education, an Hachette UK company, Market Place, Deddington, Oxfordshire OX15 0SE

Orders

Bookpoint Ltd, 130 Milton Park, Abingdon, Oxfordshire OX14 4SB
tel: 01235 827720
fax: 01235 400454
e-mail: uk.orders@bookpoint.co.uk

Lines are open 9.00 a.m.–5.00 p.m., Monday to Saturday, with a 24-hour message answering service. You can also order through the Philip Allan Updates website: www.philipallan.co.uk

© Philip Allan Updates 2009

ISBN 978-0-340-95857-5

First printed 2009
Impression number   5   4   3   2   1
Year   2014   2013   2012   2011   2010   2009

Printed in Spain

Hachette UK's policy is to use papers that are natural, renewable and recyclable products and made from wood grown in sustainable forests. The logging and manufacturing processes are expected to conform to the environmental regulations of the country of origin.

P1637

Contents

Introduction vi

Topic 1   Plate tectonics, earthquakes and volcanic hazards

Unit 1

    1 Plate tectonics 1
    2 Volcanoes 3
    3 Volcanic hazards 4
    4 Earthquakes and earthquake hazards 7

Topic 2   Sub-aerial processes

    1 Weathering 13
    2 Mass movement 15

Topic 3   Fluvial environments

    1 Fluvial processes 19
    2 River channels 20
    3 Fluvial landforms 24
    4 River flooding 27

Topic 4   Cold environments

    1 Distribution of cold environments 33
    2 The chronology of glaciation 33
    3 Types of glacier 34
    4 Formation of glacier ice 34
    5 The movement of glaciers 34
    6 Mass balance 35
    7 Glacial erosion 35
    8 Glacial deposition 38
    9 Periglaciation 40
   10 Ecosystems in cold environments 43
   11 Human activities in cold environments 44

Topic 5   Coastal environments

    1 The coastal system 46
    2 Energy inputs 46
    3 Landforms of coastal deposition 47
    4 Mudflats and salt marshes 51
    5 Sand dunes 52
    6 Landforms of coastal erosion 53
    7 Rock structure and the planform of coasts 56
    8 Sea level change 56
    9 Coastal management 57

Topic 6   Hot arid and semi-arid environments

Unit 1

    1 Distribution of hot arid and semi-arid environments 61
    2 The causes of aridity 62
    3 Weathering processes 63

4 Aeolian processes and landforms.. 63

5 Fluvial processes and landforms 65

6 Human impacts in hot arid and semi-arid environments............................... 66

7 Managing desertified and degraded land... 69

Topic 7 Climate change and climate hazards

1 Climate change... 71

2 Evidence for climate change 71

3 Global warming: anthropogenic climate change 72

4 The impacts of global warming.. 74

5 Predicting climate change.. 76

6 Responding to climate change 77

7 Climatic hazards... 78

8 Hurricanes.. 78

9 Tornadoes 81

10 Extreme weather conditions 82

11 Anticyclones... 83

12 Depressions.. 84

Topic 8 Ecosystems

1 Ecosystems.. 86

2 Nutrient cycles 88

3 Ecological succession.. 89

4 The effect of human activity on natural ecosystems...................................... 90

Topic 9 Population and resources

1 Population change... 91

2 Fertility and mortality... 92

3 Age–sex structure 94

4 Migration.. 96

5 Population policies 98

6 Natural resources 99

7 Population and resources.. 100

Topic 10 Rural change and management

1 Defining rural areas 102

2 Rural change in the UK 102

3 Declining rural services... 104

4 Second homes and affordable housing... 105

5 Planning policies in rural areas 106

6 The environmental impact of rural change... 106

Topic 11 Urban change: problems and planning

1 Defining urban populations and urban areas 109

2 Urbanisation and urban growth.. 109

3 Global urbanisation 109

4 Urban growth and city size 110

5 World cities 111

6 Urban social and economic changes in developed countries.......................... 112

Unit 1

unit 2

   7 Urban social and economic changes in developing countries 114

   8 Urban inequality.. 116

   9 Urban areas and the environment 117

  10 Sustainable cities 119

Topic 12 Globalisation

   1 What is globalisation?... 122

   2 The causes of globalisation 122

   3 Transnational corporations (TNCs) 124

   4 TNCs and social and economic issues... 125

   5 TNCs, globalisation and the environment.. 127

Topic 13 Development

   1 The development gap 129

   2 Measuring development 130

   3 Explaining the development gap 131

   4 Dependency theory.. 134

   5 Reducing the development gap 135

Topic 14 Energy

   1 Sources of energy 140

   2 Energy use and economic development 141

   3 Energy security 142

   4 Sustainable energy supplies.. 146

Topic 15 Tourism

   1 Defining tourism 149

   2 The environmental impact of tourism 149

   3 Sustainable tourism 150

   4 Management and planning for sustainable tourism... 151

Topic 16 Food and water

   1 Food systems 154

   2 Patterns of global food consumption.. 154

   3 Food security 155

   4 Food supplies and population growth.. 156

   5 Increasing food production: the technological 'fix' 157

   6 Water supply 160

   7 Water demand 161

   8 Development of water resources and its impact 163

   9 Sustainable water management 164

  10 Transboundary water disputes.. 165

Topic 17 Pollution and health risks

   1 Defining risks to human health 167

   2 The causes of health risks 169

   3 Pollution and health risks 173

   4 Managing health risks 175

Unit 1

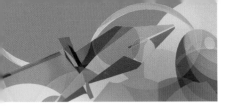

Introduction

About this book

This book provides basic knowledge and understanding of the core topics in geography at AS and A2. Two special features of this new edition are the inclusion of case studies and boxes. The case studies, though inevitably limited in detail in a book of this length, provide essential exemplification of geographical patterns, processes and changes.

A number of boxes intersperse the text. They have two purposes. Some explore topics in more depth, while others provide further exemplar material. Learning the content of the boxes is not essential, but if you require extra detail and understanding you should find them useful.

What examiners are looking for

Markers assess examination answers in geography against three criteria or *assessment objectives*: *knowledge and understanding*; *application of knowledge and understanding* and *skills*. Each unit puts a slightly different weighting on these assessment objectives. For example, a unit on geographical techniques will have a relatively high weighting for skills, whereas a synoptic unit may be weighted towards the application of knowledge and understanding.

Knowledge and understanding

Your answers will have to demonstrate both knowledge and understanding of the topics you have studied. The difference between knowledge and understanding is not always clear. As a rule of thumb, questions which test knowledge usually ask you to 'describe', 'outline', 'name' or 'state'. Those which test understanding require you to 'explain'. For example, if you *described* the shape of a cumulus cloud you would draw on your knowledge. But to *explain* the shape of a cumulus cloud would require understanding of physical processes, such as condensation and adiabatic cooling. Explanation is a higher-order skill than description and therefore carries higher marks.

Application

To do well at AS/A2 your answers must not only demonstrate good knowledge and understanding but also show that you can apply this knowledge and understanding relevantly to a question. What often sets apart A*/A-grade candidates from the rest is their ability to apply knowledge and understanding successfully. Consider the question: why do rates of coastal erosion vary from one stretch of coast to another? An answer might show knowledge of erosion, erosion rates and different stretches of coastline. It might also show understanding of the processes which influence erosion rates (e.g. wave energy, lithology, beaches). However, application of knowledge and understanding requires you to explain how different bundles of processes operate at different locations, and thus give rise to variable rates of erosion.

Skills

The appropriate way to assess many geographical skills is through fieldwork and research. However, some skills can be assessed in written papers. Written answers subsume both literacy and numeracy skills (including quality of English). Literacy skills include the ability to plan organised answers, use geographical terminology, synthesise and write a conclusion etc. In a unit on geographical techniques you will have to demonstrate skills such as data collection, presentation and analysis, as well as ICT skills that allow access to geographical information systems and large databases operated by government departments and multilateral agencies like the United Nations (UN), World Health Organization (WHO) and World Bank (WB).

Types of question in written examinations

Examiners use a range of question types to assess knowledge, understanding and skills. It is important that you know the type of question used in assessing each unit. This information is given in the full specification, which is available on the awarding body's website. The main types of question are:

- short answer: data-response and stimulus-response questions
- structured essays divided into several sub-sections
- open-ended discursive/evaluative essays

When revising topics assessed by short-answer questions, you should appreciate the need for concise and accurate knowledge and understanding. Although detailed examples and case studies have limited importance for short-answer questions, they are essential for essay-type questions. Learning how to plan and structure essays is also a vital component of exam practice and revision.

Responding to command words and phrases

Examination questions spell out what you must do with simple command words and command phrases. These commands must be interpreted accurately and must be followed explicitly. Because they vary in their level of difficulty, they are the principal way in which examiners achieve differentiation in their marking. Too often candidates ignore the more demanding command words and phrases such as 'discuss', 'evaluate' and 'to what extent' and instead resort to description or narrative. Study the command words in Table 1 and make absolutely sure that you understand what each one is asking you to do.

Table 1 Common command words and their meanings

Common command words	Meaning
Compare	Describe the similarities and differences of at least two features, patterns and processes.
Describe	Description is the simplest skill and carries the fewest marks. It requires you to provide a word picture of a feature, pattern or process. Descriptions should be precise, accurate and structured. 'Outline' is an alternative command to 'describe' but requires less detail.
Examine	Describe and comment on a pattern, process or idea. 'Examine' often refers to ideas or arguments which demand scrutiny from different viewpoints.
Explain/Why...?	Explanation demands understanding as well as knowledge. Explanation is a higher-level skill than description, requiring you to understand processes and causes.
Discuss/assess/evaluate	These skills are of the highest order and are likely to be tested at A2. To discuss, assess and evaluate you are required to consider different viewpoints, outcomes or strategies and reach some overall synthesis, conclusion or reasoned judgement.
Amplify	The command word 'amplify' is asking you to describe and explain something in more detail. Some of the extra detail will probably include case studies and examples.

In your revision you should work closely with past papers. You should identify the key commands and phrases used in questions. Practise your response to these key words and phrases by preparing planned answers to past questions.

How to revise

Your revision of a topic should follow a logical sequence:

- Choose the revision topic and read the section of your specification which deals with that topic.
- Compile a list of past examination questions on the topic and identify the main themes.
- Get copies of mark schemes for these questions (download past and specimen papers and mark

schemes from the awarding body's website) so that you understand how the questions are marked.
- Read and learn the relevant notes in this textbook using the themes of past questions and examiners' mark schemes to structure your learning.
- Learn appropriate examples and case studies from the following sources: this book, other texts recommended by your teacher, articles in publications such as *Geography Review* and *Topic Eye*, publishers' CDs and online resources (www.hodderplus.co.uk/philipallan), class notes and hand-outs.
- Write and/or prepare planned answers to all past examination questions on the topic.

Structuring your revision

Most geography topics at AS/A2 have three dimensions:
- the main features of geographical forms, patterns and processes, including their variations in space and over time
- the impact of geographical processes on the human and/or physical environment and arising issues
- the human response to and management of the issues and impacts

You should adapt your revision of topics to fit this structure. An example is given in Table 2.

Table 2 Structuring revision topics: the example of land degradation

Dimensions	Example: land degradation
Main characteristics including descriptions of form, causes, variations in space and time	Soil erosion, salinisation, deforestation. Physical and human causes of soil erosion, e.g. drought, climate change, overpopulation
Impact on the human and/or physical environment	Loss of soil cover, waterlogging and salinisation of soils, local climate change, lowering of water tables etc.
Human responses	Reafforestation, land drainage, terraces, contour ploughing, maintaining soil fertility, shelter belts, land abandonment etc.

Points to remember

Throughout your revision you must:
- Have a thorough systematic knowledge and understanding of the subject.
- Learn a range of examples and case studies at different scales to illustrate your answers.
- Adopt an intelligent approach to learning, building your revision around a framework of past examination questions and mark schemes.
- Apply knowledge and understanding so that your answers reflect accurately (and therefore relevantly) the demands of the questions.
- Practise writing answers under timed conditions and prepare planned answers for those that you do not wish to write out in full.

In the examination you must:
- Plan and give structure to your answers — essay-type questions will require 5 minutes of planning to organise and specify content (factual content, arguments and exemplification).
- Manage your time efficiently — remember that if you spent excessive time on one question, the extra marks you gain are unlikely to compensate for the loss of time (and marks) on other questions.
- Convey knowledge and understanding through sketch maps, diagrams and charts, as well as text.
- Avoid writing generalised answers to essay-type questions — geography is about the real world and this should be reflected through your use of examples and case studies.
- Write answers that show detailed knowledge and understanding and a clear ability to apply knowledge and understanding accurately to the questions set.
- Make a conscious effort to write clear English, with accurate punctuation, grammar and spelling.
- Use appropriate geographical terminology accurately.

Michael Raw

TOPIC 1 Plate tectonics, earthquakes and volcanic hazards

1 *Plate tectonics*

The Earth's crust and lithosphere are broken into seven large slabs and a dozen or more minor ones known as **tectonic plates**. The distribution of the main plates is shown in Figure 1.1. The plate boundaries are zones of great tectonic activity, including volcanism, earthquakes, mountain building and faulting and folding. The global distribution of active volcanoes and earthquakes defines the plate boundaries. We recognise three types of plate boundary: constructive, destructive and conservative.

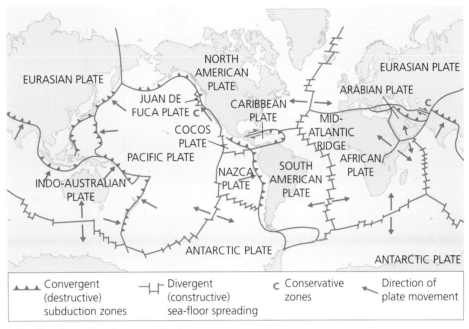

Figure 1.1 Distribution of main tectonic plates

1.1 Constructive plate boundaries

Constructive (or divergent) plate boundaries are the **mid-ocean ridges** where new crust forms. There, as Figure 1.2a shows, rising plumes of magma from the Earth's mantle stretch the crust and lithosphere. Active volcanoes develop where lava reaches the surface. Most volcanic activity takes place on the ocean floor and forms the submarine mountain ranges of the mid-ocean ridges. Parallel **faults** associated with tension in the crust and volcanism cause **rifting** and deep valleys between the mountain chains of the mid-ocean ridges.

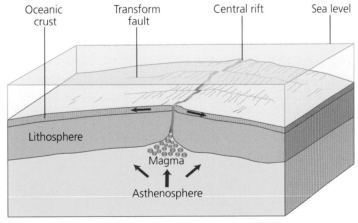

(a) A divergent plate boundary and spreading centre

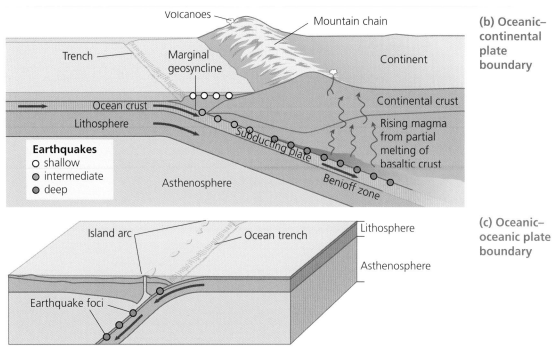

Figure 1.2 Constructive plate boundaries

1.2 Destructive plate boundaries

At **destructive** (convergent) plate boundaries or **subduction zones**, oceanic crust/lithosphere is destroyed (Figure 1.2). Subduction involves the following stages:

- the descent of the subducted plate together with water and seafloor sediments into the upper mantle
- melting of the subducted slab and surrounding mantle rocks around 100 km or so below the surface
- the melt (or magma), which is less dense than the surrounding rocks, rising slowly through lines of weakness and fissures to the surface
- the eruption of lava, gases and ash at the surface through volcanoes and fissures

Subduction occurs when two tectonic plates converge. The older, denser plate is subducted. If two oceanic plates converge (Figure 1.2c), subduction forms an island arc such as the Kuril Islands in the north Pacific and the Lesser Antilles in the Caribbean.

The subduction of an oceanic plate beneath a continental plate produces fold mountain chains such as the Andes and the Cascades along the Pacific coast of the Americas (Figure 1.2b). Destructive plate boundaries are also the location of volcanoes, earthquakes and ocean trenches.

Fold mountain chains

The world's highest mountain ranges, including the Himalayas, Andes and Alps, are located along subduction zones.

Where an oceanic plate and continental plate converge, an island arc may collide with the continent and contribute to mountain building. Meanwhile, sedimentary rocks formed on the continental shelf and continental slope, squeezed between the island arc and the continent, crumple to form mountain ranges. Subduction of the oceanic plate may produce huge **intrusions** of magma beneath the mountains, which creates further uplift. This sequence of events explains the formation of the Andes in South America.

The Himalayas have been formed by the convergence of two continental plates: the Indo-Australian plate and the Eurasian plate. As the two plates converged, the Tethys Sea narrowed until its sea-floor sediments were pushed nearly 9 km above sea level into complex folds. In south Asia two continental landmasses (Eurasia and India) have collided. The collision has welded the continents together, producing a great thickness of the continental crust. As a result, there is no volcanic activity in the Himalayas.

Ocean trenches

Narrow trenches, hundreds of kilometres long and up to 11 km deep, occur on the ocean floor parallel to island arcs and fold mountain ranges. Ocean trenches mark the zone of subduction, where oceanic crust/lithosphere descends into the mantle. As it does so, the leading edge of the overriding plate is buckled to form a trench.

1.3 Conservative plate boundaries

At **conservative** plate boundaries, two plates slide past each other with a shearing movement. This movement can be violent and can cause severe earthquakes. However, volcanism is absent. In southern and central California, the boundary of the Pacific and North American plates forms a conservative plate margin known as the San Andreas fault. Earthquakes occur frequently along this fault line and present major hazards to metropolitan areas such as San Francisco and Los Angeles.

2 *Volcanoes*

2.1 Volcanoes

A volcano is an opening in the Earth's crust where molten rock and gases reach the surface. The **ejecta** or fragments thrown out by an eruption include lava, pumice, cinders, ash and gases. The nature of these materials is variable and explains differences in the shape of volcanoes and the nature of volcanic eruptions.

Many volcanoes such as Mt Fuji have a classic conical shape, as shown in Figure 1.3. These **strato-volcanoes** comprise layers of lava, ash and other ejecta erupted by the volcano. The **vent** occupies a collapsed hollow which, depending on its size, is known as a **crater** or **caldera**. Feeding the volcano and located 3 or 4 km underground is the **magma chamber**. Magma from the mantle fills this chamber before an eruption. The build-up of magma is detectable at the surface because the ground swells or inflates. Inflation can tear the crust apart to form rifts or fissures at the surface. Fissure eruptions are common in Iceland and Hawaii.

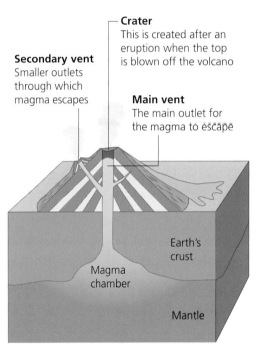

Crater
This is created after an eruption when the top is blown off the volcano

Secondary vent
Smaller outlets through which magma escapes

Main vent
The main outlet for the magma to escape

Earth's crust

Magma chamber

Mantle

Figure 1.3 A strato-volcano

2.2 Types of volcano

Hawaiian-type volcanoes

The Hawaiian Islands in the Pacific Ocean are one of the most active volcanic areas in the world. Located at the centre of the Pacific plate, volcanic activity is related to neither subduction nor a spreading ridge but to a **hot spot**, a rising mantle plume that has punched a hole through the crust.

This hot spot has remained fixed in position for over 70 million years. However, during this time the Pacific plate has moved (at a speed of just 3 or 4 cm a year) in a northwesterly direction over the hot spot. Today, the Hawaiian hot spot is over the Big Island and it is here that active volcanism is found. The Big Island has formed in the last 1 million years by eruptions from its five volcanoes. The largest — Mauna Kea and Mauna Loa — reach over 4,000 m above sea level and rise 9,000 m from the ocean floor. Kilauea is the most active volcano: it has been in continuous eruption since 1983.

The Loihi seamount, located 35 km off the southeast coast of the Big Island, is the youngest volcano in the Hawaiian chain. Rising 3 km above the ocean floor, its summit should break the surface in 10,000–100,000 years.

In profile the Big Island's volcanoes have the shape of a flattened dome. They are known as **shield volcanoes**. They are giants — at its base Mauna Loa is 120 km in diameter. However, its slopes never exceed a gentle 12°. Hawaii's shield volcanoes form for the following reasons:

● Most eruptions consist of lava rather than ash and gas.
● The basalt lava has only a relatively small proportion of silica. It is non-viscous, has a low gas content, and flows for long distances before cooling and solidifying.
● Eruptions are relatively gentle with little explosive activity.

Strato-volcanoes

Strato-volcanoes consist of layers of ash and lava. They have steeper slopes than shield volcanoes and a more conical shape. In contrast to shield volcanoes, the magma that forms strato-volcanoes is viscous and has a high silica content. It is known as **andesite**. Viscous magmas like andesite create explosive products such as cinder and ash, and fewer lava flows. In viscous magma, trapped gases such as steam cannot escape easily. As a result, the pressures caused by the build-up of gases can lead to devastating eruptions which can blow a volcano apart.

This is precisely what happened at Mt St Helens in the Cascade range of the northwest USA in May 1980. A massive explosion blew away the top 400 m of the volcano. Ash and pyroclasts combined with hot gases to form **pyroclastic flows** that destroyed everything in their path. Melting snow mixed with ash caused destructive debris flows and **lahars**. The result was total devastation within a 15 km radius of the volcano.

3 Volcanic hazards

3.1 Natural hazards

Volcanic eruptions, together with earthquakes, mass movements, hurricanes, tornadoes and floods, are **natural events** that may become **natural hazards**. Natural events that are harmful to people, causing death, injury and damage to property and infrastructure, are known as natural hazards. Large-scale natural hazards that result in major loss of life and widespread damage are called **natural disasters**.

The impact of natural hazards can be explained by two concepts:
● exposure, i.e. the size or scale of the natural event and the number of people in the area affected
● vulnerability, i.e. the preparedness of a country or population to cope with a hazard and the economic status of the affected population

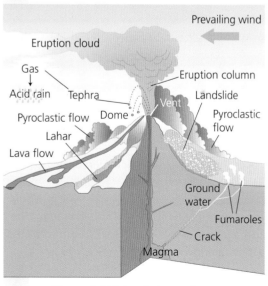

Figure 1.4 Volcanic hazards

3.2 The variable impact of eruptions

Volcanic eruptions produce a range of hazards (Figure 1.4). Their impact on people depends on three factors.

● The nature of the volcanic ejecta (i.e. lava, tephra, gas) and their violence. Gentle, effusive eruptions of lava, such as those found in Hawaii, pose little direct threat to human life. Even so, lava flows destroy farmland, buildings and infrastructure. Explosive eruptions of superheated gas and tephra cause total devastation.
● The density of population living in the vicinity of the volcano.
● Monitoring and warning systems and evacuation procedures for the population at risk. Compared to developing countries, developed countries have sophisticated monitoring, early-warning and evacuation procedures. As a result, loss of life in developed countries is greatly reduced.

3.3 Types of volcanic hazard

The main types of volcanic hazard are shown in Figure 1.4.

Lava flows

The flow of **lava** is almost unstoppable, although lava flows have occasionally been halted by cooling them with water (at Heimaey, Iceland in 1974 — see Table 1.1) and diverted by explosions (e.g. Mt Etna, Sicily). Lava flows, although rarely a threat to life, can cause enormous damage to property. In 1990 a lava flow from Puu Oo crater in Hawaii buried the village of Kalapana. Further eruptions between 1992 and 1993 destroyed 181 homes, buried large areas of farmland and severed the main coastal road.

Pyroclastic flows

Pyroclastic flows are high-speed avalanches of hot ash, rock fragments and gas that destroy everything in their path. They can reach speeds of 200 km/h and temperatures of over 1,000°C.

Soufrière Hills volcano on the Caribbean island of Montserrat, which erupted between 1995 and 1998, produced many pyroclastic flows (Case Study 1, page 6). These flows destroyed much of the island's farmland, led to the evacuation of half of Montserrat's population of 11,000 and the abandonment of the southern half of the island, including the capital, Plymouth.

Lahars

Lahars are mixtures of water, rock, sand and mud that flow down valleys leading away from a volcano. They can be caused by:

- an eruption melting snow and icefields around a volcano's summit
- the rapid release of water following the breakout of a summit crater lake
- heavy rainfall washing away loose volcanic ash

Lahars are fast-moving and can travel long distances. They are particularly destructive because they follow valleys where settlements and population are often concentrated. One of the largest lahar events occurred on Mt Rainier in northwest USA about 5,700 years ago. Rocks and mud swept down the White Valley to reach their present position near Tacoma, 120 km from the volcano.

Jökulhlaups

Even more catastrophic are volcanic eruptions beneath an icefield or glacier. Rapid melting of ice releases enormous volumes of water and generates massive floods. In Iceland these floods are known as **jökulhlaups**. Iceland's most recent jökulhlaup occurred in 1996 following the eruption of the Grímsvötn volcano beneath the Vatnajökull icefield. A peak flow of 45,000 cumecs was recorded. The flood, which lasted for a week, destroyed several bridges and 10 km of the ring road that encircles the island. However, because the flood was expected, dykes were strengthened and people were evacuated. There was therefore no loss of life or damage to settlements.

Climate change

Large-scale explosive eruptions may affect the global climate. Following a major eruption, droplets of sulphuric acid and dust often remain suspended in the atmosphere for several years. These particles reflect and absorb **insolation** lowering temperatures at the surface. The eruption at Mt Pinatubo (Philippines) in 1991 caused a significant cooling of the global climate in the following year. The eruption of Mt Tambora (Indonesia) in 1815 was responsible for one of the coldest summers on record. Crops failed worldwide and famine caused millions of deaths.

3.4 Mitigating volcanic hazards

Mitigation of volcanic hazards depends on monitoring and warning people of impending eruptions (Table 1.1). Monitoring includes recording seismic shocks, measuring ground inflation and collecting gas and lava samples. **Hazard mapping** can reveal areas most at risk from lava flows, lahars and

pyroclastic flows. Preparedness is most advanced in developed countries such as the USA and Japan, where it greatly reduces risks and loss of life.

Table 1.1 Mitigating volcanic hazards

Scheme	Description
Monitoring	Earthquakes and tremors develop as magma forces its way to the surface inside the volcano. These shocks are recorded by **seismometers** on the volcano. Gravity is also measured: as magma fills the reservoir beneath the volcano, gravity increases. Gases are sampled. Rising levels of sulphur dioxide and hydrogen chloride signal an impending eruption. Ground deformation (inflation) as magma accumulates within the volcano is further evidence of an imminent eruption.
Diversion of lava	Small lava flows have been successfully diverted away from centres of population. At Heimaey in Iceland, the fishing harbour was saved by spraying a lava flow with sea water.
Hazard mapping	The paths followed by ancient lahars and pyroclastic flows can be mapped from sediments.
Warning and evacuation	Lahar detection warning systems have been installed around Mt Rainier in Washington state. Detection triggers an automatic alert that initiates evacuation.

Box 1   *Benefits of volcanoes*

Although volcanic eruptions often cause death and destruction, they can also benefit people.

■ Volcanic ash and lava are rich in minerals and form fertile soils. The high density rural populations in Java, Indonesia are largely supported by intensive farming of rich volcanic soils.

■ Lava flows may create new areas of land. The lava flows generated by the Kilauea volcano in Hawaii since 1983 have added 200 ha to the area of the Big Island.

■ Volcanoes often attract visitors and thus help local economies. The Volcanoes National Park in Hawaii attracts nearly 2 million visitors a year, while Yellowstone National Park in the USA, with its famous geysers, has over 4 million visitors a year. The world's two most climbed mountains are both volcanoes: Mt Fuji in Japan and Mt St Helens in the USA.

■ In Iceland and New Zealand volcanic activity is an important source of geothermal energy. Hot water from volcanism provides central heating for Iceland's capital, Reykjavik. Elsewhere, pumice and ash deposits are used by the construction industry.

CASE STUDY 1	Soufrière Hills, Montserrat, 1995–2003
Cause	Montserrat is part of an island arc, formed by the subduction of the North American plate below the Caribbean plate. Montserrat owes its existence to the Soufrière Hills strato-volcano.
Hazards	Pyroclastic flows, ash falls, debris avalanches and occasional lava flows. The volcano is explosive; its magma is thick, viscous and andesitic.
Exposure	Soufrière Hills' eruptions are explosive and deadly pyroclastic flows carry high levels of risk. Given the small size of the island, Montserrat is densely populated. In 1990 nearly 11,000 people lived in areas at risk from pyroclastic flows and ashfalls.
Vulnerability	Vulnerability was high because the eruptions occurred on a small island. However, this vulnerability was reduced by close monitoring of the volcano. Using data on seismic activity, volcanic gases and ground deformation, scientists have been able to issue early warnings and prepare people for evacuation. In the early eruptive stages, the area most at risk (i.e. the southern half of the island and the capital, Plymouth) were evacuated and designated an exclusion zone.
Impact	Eruptive activity peaked in 1997 when 19 islanders were killed and Plymouth was destroyed by pyroclastic flows, ashfalls and fires. Between 1990 and 2000 Montserrat's population fell from 11,000 to 6,500. Today, two-thirds of the island is uninhabitable and most fertile farmland in the south has been destroyed. Tourism, the former mainstay of the economy, has been ruined.

CASE STUDY 2	Nyiragongo, Democratic Republic of Congo, 2002
Cause	Nyiragongo is a strato-volcano in East Africa's rift valley. At this divergent plate boundary, volcanism is due to tension and stretching of the continental crust and lithosphere.
Hazards	Lava flows and toxic gases (especially sulphur dioxide). The magma is low in silica, fast-moving and a major threat to life, property, farmland and livestock.
Exposure	Exposure was high because: • Nyiragongo is one of Africa's most active and deadly volcanoes • 500,000 people live in the vicinity of the volcano
Vulnerability	Vulnerability was high. The local population is poor, depends heavily on subsistence farming and has few resources to buffer it against eruptions. Recent civil wars have increased poverty and vulnerability. Education of local people to raise awareness of risks and early warning and evacuation procedures are needed to reduce vulnerability.
Impact	The Nyiragongo crater contains an active lava lake, which can drain suddenly. This happened in 2002, generating lava which flowed into the city of Goma. Fourteen villages and one-fifth of Goma were destroyed, 110 people died and 120,000 were made homeless. Four-fifths of the local economy was destroyed.

4 *Earthquakes and earthquake hazards*

4.1 Causes of earthquakes

Earthquakes are vibrations (**seismic waves**) in the Earth's crust caused by the fracturing of rocks and sudden movements along fault lines. They result in violent shaking of the ground (the **primary hazard**), **liquefaction**, **landslides** and **tsunamis** (**secondary hazards**). The world's major earthquake zones correspond to plate boundaries. **Inter-plate** movements cause tension, compression and shearing of the crust. Rocks under pressure may eventually snap, resulting in crustal movements along fault lines that release huge amounts of energy as seismic waves. Major earthquakes also occur thousands of kilometres away from plate boundaries. Recent **intra-plate** quakes, caused by slippage along fault lines, include those in Gujarat (2001) and Sichuan (2008).

The precise location of an earthquake within the crust is known as the **focus**. The point on the surface immediately above the focus is the **epicentre**. The destructive power of an earthquake is greatest close to the epicentre. Earthquakes of similar magnitude are more destructive if they occur near the surface.

Box 2 *Seismic waves*

Earthquakes produce two principal types of seismic wave: P-waves and S-waves. In the Earth's crust, P-waves travel at around 6–7 km s⁻¹, while S-waves travel more slowly (2.5–4 km s⁻¹). P-waves, like sound waves, consist of successive compression and stretching of particles in the rocks (Figure 1.5). The motion of these particles is parallel to the direction of the wave. P-waves travel through both solids and liquids. S-waves are transverse waves, which means that particle motion is sideways. S-waves cannot travel through liquids.

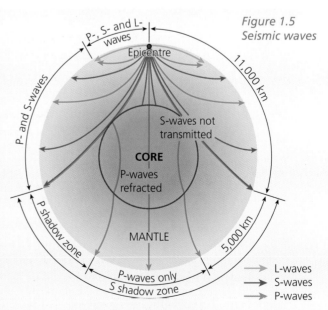

Figure 1.5 Seismic waves

4.2 Earthquake magnitude and intensity

The **Richter scale** measures earthquake magnitude. Earthquakes range in magnitude from 2.5 to 9 on the Richter scale. **Seismographs** record the amplitude of earthquake waves, which radiate in all directions from the focus. These waves give a measure of the amount of energy released by an earthquake. The Richter scale is determined by the logarithm of the amplitude of seismic waves. In terms of energy release, a magnitude 7 quake is around 30 times more powerful than a magnitude 6 quake and 900 times greater than a magnitude 5 event.

The **Mercalli scale** measures earthquake intensity, i.e. the impact of an earthquake on people and structures. The scale goes from 1 to 12, where 1 is instrumental (i.e. detected only by seismographs) and 12 is catastrophic, causing total destruction.

There is little relationship between earthquake magnitude and intensity. For example, a magnitude 6.7 quake struck Los Angeles in 1994 and killed 57 people. Four months earlier an earthquake of similar magnitude in central India caused 22,000 deaths.

Box 3 *The San Andreas fault*

The San Andreas fault in southern and central California is a conservative plate boundary (Figure 1.6). It separates the Pacific and North American plates and is one of the most active earthquake zones in the world. The Pacific plate is sliding northwest at a speed of a few centimetres a year. However, this movement is not smooth. Friction between the plates restricts movement, causing pressure to build up. When movement occurs there is a sudden release of stored energy, which is an earthquake. Major earthquakes associated with the San Andreas fault occur every 20 or 30 years. The great 1906 quake (magnitude 8.1), which destroyed San Francisco and killed around 700 people, was one of the most powerful ever recorded. San Francisco was hit by another large quake in 1989, though damage was less severe. The Northridge quake near Los Angeles in 1994 killed 57 people and caused damage estimated at US$20 billion (Case Study 4, pages 11–12).

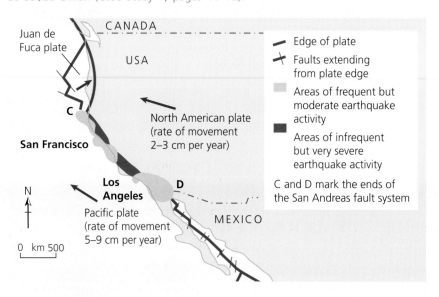

Figure 1.6 San Andreas fault system

4.3 Social, economic and environmental impacts

Earthquakes damage buildings and infrastructure, and cause injuries and death. Large earthquakes can devastate an entire region and kill or injure thousands of people. Collapsed buildings and other structures are the main cause of death and injury. In the aftermath of an earthquake, fire and disease may add significantly to the death toll.

More than one-third of the world's largest cities (most of them in developing countries) are located in active seismic zones.

The risk to people from an earthquake (and other natural hazards) can be summarised by the following formula:

$$R = \frac{M \times P}{V}$$

where: R = risk
  M = size/scale of the earthquake
  P = number of people living in the affected area
  V = vulnerability (i.e. preparedness — building regulations, disaster planning, education, level of development)

In other words, risk is directly proportional to the size/scale of the event and the number of people living in the immediate area, and is inversely proportional to the preparedness of society to counter the hazard.

People who live in earthquake zones can do nothing to mitigate the magnitude of a quake and are unlikely to reduce population densities. An earthquake that strikes at night, when most people are asleep indoors, will cause more death and injury than a daytime quake. It is also clear that the more densely populated a region is, the more people are at risk from earthquakes.

Earthquakes are more damaging in poor countries, which lack the resources (a) to construct earthquake-proof buildings and other structures, and (b) to put in place effective emergency procedures to deal with disasters quickly and effectively.

4.4 Mitigating earthquake impacts

In contrast with other natural hazards, it is impossible to give early warning of earthquakes with any accuracy. Because earthquakes occur suddenly and unexpectedly, this makes them particularly deadly. The main human response to earthquake hazards is to minimise (or mitigate) their impact. However, we do know that in active earthquake zones such as California and Tokyo Bay, the longer the interval without an earthquake, the higher the probability of occurrence and the greater its magnitude is likely to be.

Building design

Building technology is controllable and is a significant influence on the amount of damage, death and injury caused by earthquakes. In poor countries, few buildings are earthquake-proof. In rural areas in developing countries, traditional houses with heavy roof timbers and mud walls collapse easily, trapping their occupants. In urban areas, multi-storey flats and reinforced concrete buildings — often built cheaply and with safety standards ignored — may collapse, leading to high death tolls. Although many developing countries have strict building codes, as revealed in the Sichuan (2008) and Kashmir (2005) quakes, these codes are rarely enforced rigorously (Case Study 3, page 10).

Rich countries such as Japan and the USA, which straddle active earthquake zones, may avoid building high-rise structures in areas most at risk. However, in densely populated urban centres such as Tokyo or San Francisco this may not be an option. In these areas, strict building regulations are enforced to ensure that buildings and other structures are earthquake-proof.

Earthquake-proof high-rise buildings include designs with:
● steel frames and braces that twist and sway during an earthquake without collapsing
● foundations mounted on rubber shock absorbers
● deep foundations into the bedrock
● first-storey car parks allowing the upper floors to sink and cushion the impact
● concrete counter-weights on the top of buildings, which move in the opposite direction to the force of the quake

In developing countries, earthquake disasters are most linked to poor building construction and design. For example, in Gujarat (2001), Bam (2003), Kashmir (2005) and Sichuan (2008), most deaths and injuries were caused by building collapse.

CASE STUDY 3   Kashmir, Pakistan, 2005

| Physical details |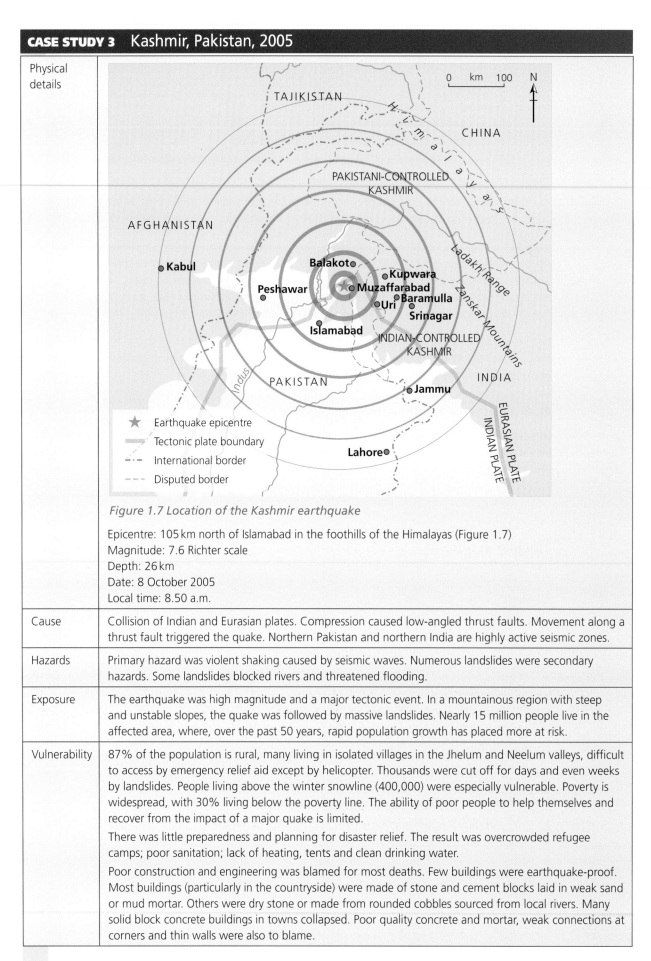

Figure 1.7 Location of the Kashmir earthquake

Epicentre: 105 km north of Islamabad in the foothills of the Himalayas (Figure 1.7)
Magnitude: 7.6 Richter scale
Depth: 26 km
Date: 8 October 2005
Local time: 8.50 a.m. |
| --- | --- |
| Cause | Collision of Indian and Eurasian plates. Compression caused low-angled thrust faults. Movement along a thrust fault triggered the quake. Northern Pakistan and northern India are highly active seismic zones. |
| Hazards | Primary hazard was violent shaking caused by seismic waves. Numerous landslides were secondary hazards. Some landslides blocked rivers and threatened flooding. |
| Exposure | The earthquake was high magnitude and a major tectonic event. In a mountainous region with steep and unstable slopes, the quake was followed by massive landslides. Nearly 15 million people live in the affected area, where, over the past 50 years, rapid population growth has placed more at risk. |
| Vulnerability | 87% of the population is rural, many living in isolated villages in the Jhelum and Neelum valleys, difficult to access by emergency relief aid except by helicopter. Thousands were cut off for days and even weeks by landslides. People living above the winter snowline (400,000) were especially vulnerable. Poverty is widespread, with 30% living below the poverty line. The ability of poor people to help themselves and recover from the impact of a major quake is limited.

There was little preparedness and planning for disaster relief. The result was overcrowded refugee camps; poor sanitation; lack of heating, tents and clean drinking water.

Poor construction and engineering was blamed for most deaths. Few buildings were earthquake-proof. Most buildings (particularly in the countryside) were made of stone and cement blocks laid in weak sand or mud mortar. Others were dry stone or made from rounded cobbles sourced from local rivers. Many solid block concrete buildings in towns collapsed. Poor quality concrete and mortar, weak connections at corners and thin walls were also to blame. |

Impact	87,000 people died and 3 million were made homeless. Most deaths were due to collapsed buildings. There were further deaths after the quake due to injury, exposure and disease. Many survivors were forced to sleep outside and cope with severe winter weather in the mountains. Half of all the buildings in Muzaffarabad were destroyed. There was extensive damage to roads, bridges, schools, hospitals and electricity infrastructure. Four-fifths of crops and half of all arable land were destroyed. One hundred thousand cattle were killed. Total cost of the quake: US$5 billion.

Box 4   *Tsunamis*

Tsunamis are huge waves at sea, usually caused by earthquakes. In the open ocean, tsunamis may be less than 1 m high and pass unnoticed. However, as they approach shallow coastal waters, wave heights can increase dramatically (8–15 m) and overwhelm coastal settlements.

Tsunamis travel at speeds of up to 800 km/h. Early warning of tsunamis in coastal areas close to an earthquake epicentre is almost impossible. However, if a tsunami forms thousands of kilometres away on the opposite sides of an ocean, the authorities can give early warnings and evacuate areas at risk. Around the Pacific Ocean a tsunami warning system is operational. However, a similar system did not exist in the Indian Ocean when the 2004 Boxing Day tsunami devastated the west coast of Sumatra (Indonesia) and parts of Thailand, causing 280,000 deaths. This tsunami was triggered by a 9.1 magnitude earthquake in the subduction zone off the west coast of Sumatra.

In May 1960 a tsunami caused by an earthquake off the coast of Chile struck the town of Hilo in Hawaii. It killed 61 people and caused damage to property of over US$20 million. The wave completely destroyed areas of the town fronting the Pacific Ocean. Instead of being rebuilt, the low-lying coastal strip was turned into parks and the survivors were relocated to higher ground. Apart from early warning, other mitigating actions against tsunami hazards are:

■ increasing public awareness (e.g. publication of maps showing susceptible area, safety zones and direct routes to high ground; practising evacuation drills in high-risk areas)
■ construction of tsunami shelters

Disaster planning and prevention

In Japan, cities in earthquake zones have disaster plans to manage major earthquake events. The Tokyo Metropolitan Government is responsible for earthquake planning in the capital and aims to make the city 'disaster-resistant'. This involves upgrading millions of houses to make them fireproof; strengthening roads, expressways, bridges and public buildings; planning for evacuation to safe locations such as city parks (23 refuge sites in all); designating over 3,000 public shelters to house 4.25 million people in an emergency; and educating people in disaster awareness and how to cooperate with other citizens to build a 'strong society' against earthquakes.

CASE STUDY 4   Northridge, Los Angeles, 1994

Physical details	Epicentre: Northridge in the San Fernando valley, north of Los Angeles in southern California Magnitude: 6.7 Richter scale Depth: 17.5 km Date: 17 January 1994 Local time: 4.30 a.m. The sediment-filled San Fernando valley and adjacent mountains comprise east–west trending structures.	

Figure 1.8 Location of the Northridge earthquake

Cause	Movement occurred along a previously unknown (blind) thrust fault. Faulting, caused by convergence between the Big Bend of the San Andreas fault and the northwest motion of the Pacific plate, is responsible for numerous faults (Figure 1.8).
Hazards	10–20 seconds of strong shaking, which damaged buildings and infrastructure. The quake caused thousands of landslides and other slope failures.
Exposure	The Los Angeles metropolitan area has a population of 16.5 million and an average density of 2,500 persons/km^2. The Los Angeles basin is one of the most seismically active in the USA. Exposure was therefore high.
Vulnerability	California is the richest state in the world's richest country. Massive investment has helped to reduce vulnerability despite high levels of exposure. The environment of southern California is designed for seismic resistance. There are stringent building codes, high levels of preparedness and emergency response procedures in place.
Impact	The shallow focus of the quake and the densely populated, built-up Los Angeles basin account for the massive damage that resulted. The death toll was 57, and 9,000 people were injured and 20,000 displaced from their homes. Most of the damage was concentrated in the San Fernando and Simi valleys. The economic cost of the quake was US$20 billion. Motorways collapsed at seven sites and 170 bridges sustained damage. The collapse of the I-5/SH-14 interchange near San Fernando was one of the costliest failures. Near the epicentre, well-engineered buildings survived the shaking without damage. Elsewhere there were many structural failures which pointed to deficiencies in design and construction methods. Steel-framed buildings (including schools and hospitals) cracked and reinforced concrete columns were crushed. Several low-rise apartment buildings, constructed above open-air parking spaces, collapsed. Investigations following the quake showed a need to improve building codes.

TOPIC 2 Sub-aerial processes

Sub-aerial processes include weathering and mass movements. These processes are quite separate and distinct from erosion, transport and deposition by rivers, glaciers, waves and wind.

1 Weathering

Weathering is the *in situ* breakdown of rocks at or near the Earth's surface by chemical, physical and biological processes. Once the rocks have been broken down, the process of erosion transports the weathered materials to new locations. The type of weathering and its effectiveness depend on lithology and climate.

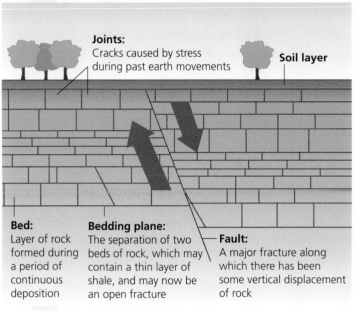

Joints:
Cracks caused by stress during past earth movements

Soil layer

Bed:
Layer of rock formed during a period of continuous deposition

Bedding plane:
The separation of two beds of rock, which may contain a thin layer of shale, and may now be an open fracture

Fault:
A major fracture along which there has been some vertical displacement of rock

Figure 2.1 Block disintegration

1.1 Products of weathering

Weathering produces rock particles and minerals that vary from fine clays to boulders. In the humid tropics, the breakdown of silicate minerals such as feldspar often produces large quantities of clay.

Meanwhile, intense chemical weathering below the surface attacks the corners and edges of rocks to form rounded boulders or **corestones**. This is an example of **spheroidal weathering**.

Granular disintegration describes weathered rock particles just a few millimetres in diameter. This is a typical product of weathered sandstone, where weathering has destroyed the cements that normally bind the sand particles together.

When rocks break down into boulders, we refer to this as **block disintegration**. Such rocks usually have massive, widely spaced joints (Figure 2.1).

The weathered rock particles and mineral materials (including soil) form a surface layer of debris known as **regolith**.

1.2 Physical weathering

Physical or mechanical weathering breaks rocks into smaller particles. During this process, the rocks and their minerals undergo no chemical alteration. The major physical weathering processes are pressure release or dilatation, insolation weathering, salt weathering and freeze–thaw.

Physical weathering merely causes the mechanical breakdown of rocks into smaller particles — there is no chemical change — increasing the surface area of exposed rock and thus encouraging further weathering.

Pressure release or dilatation

Pressure release breaks up rocks without the intervention of the weather. When prolonged erosion removes vast thicknesses of rock (i.e. **unloading**) and exposes rocks formed at depth, these rocks expand and crack. Expansion occurs parallel to the surface to form **sheet jointing** and **pseudo-bedding planes**. These lines of weakness are easily exploited by weathering.

Pressure release caused by unloading (especially in granite) may produce smooth rounded **exfoliation domes** or **bornhardts**. The outer layers of granite peel away in a process known as **sheeting**. Classic examples of granite exfoliation domes occur in Yosemite, in the Sierra Nevada of California.

Insolation weathering

In hot arid and semi-arid environments, surface temperatures in excess of 90°C have been recorded in summer. Such temperature changes lead to the thermal expansion and contraction of rock minerals. At one time it was thought that such temperature fluctuations could cause rocks to disintegrate. Experiments have since disproved this. Rock breakdown through **insolation weathering** does take place, but only when moisture (from rain and dew) is present. Insolation weathering often leads to **exfoliation**. This is when thin outer layers of rock peel away to produce smooth, rounded surfaces. However, chemical weathering processes such as hydration and oxidation also cause exfoliation.

You should appreciate that at the global scale the importance of types of physical weathering varies with climate. Thus insolation and salt weathering predominate in hot, dry climates. Freeze–thaw is important in high latitude and high altitude climates.

Salt weathering

Salt weathering is most common in hot arid and semi-arid environments. In these dryland areas, intermittent streams and rivers transport heavy loads of mineral salts in solution. This runoff often drains to inland basins, where high temperatures and high evaporation rates lead to the formation of salt pans and dry lake beds. The salt is widely dispersed by winds. When it rains, salt in solution seeps into porous rocks. There the salts precipitate, forming salt crystals. Crystal growth sets up stresses inside the rocks (rather like freeze–thaw weathering) and eventually result in rock distintegration.

Freeze–thaw weathering

When water freezes, its volume increases by 9 %. If freezing occurs in confined spaces such as joints and rock crevices, the ice exerts enormous pressure on the surrounding rock. The forces involved are sufficient to break apart even the most resistant rocks. This is **frost wedging**.

The main influence on rates of frost wedging is the number of **freeze–thaw** cycles. Other influences are the availability of liquid water and rock type.

- Ideal climatic conditions for frost wedging are where temperatures fluctuate above and below freezing. We call these temperature changes freeze–thaw cycles. Such cycles were most frequent in the British Isles during the cold, periglacial conditions that followed the last glacial (or ice age). In the British Isles today, significant freeze–thaw weathering is confined to the mountains above 800 m.
- Rocks disintegrate more rapidly by freeze–thaw action where plenty of liquid water is available. For this reason, dry tundra areas and cold deserts may experience less freeze–thaw weathering than many temperate humid environments.
- Rocks vary in their susceptibility to freeze–thaw weathering. Tough gritstones and granites are more resistant than soft shales or porous chalk. Jointing is also important. In the mountains of Wester Ross in northwest Scotland, densely jointed Cambrian quartzite breaks up more rapidly than the more massively jointed Torridonian sandstone.

1.3 Chemical weathering

Chemical weathering breaks down rocks into their chemical constituents, or alters the chemical and mineral composition of rocks.

- Most chemical weathering processes involve water.
- Temperature controls chemical weathering: rates of chemical reaction double for every 10°C rise in temperature.

Chemical weathering is most rapid in humid tropical environments where temperatures are high throughout the year. Also, the moisture needed for most chemical weathering processes is available in abundance. Rates of chemical weathering are slow in tropical arid and semi-arid environments, where the supply of moisture is limited. Chemical weathering has little effect in cold climates, because low temperatures slow chemical reactions and water is frozen for most of the year.

The four main types of chemical weathering are **solution**, **oxidation**, **hydrolysis** and **hydration**.

Solution

Rainwater, containing carbon dioxide from the atmosphere and the soil, is a weak carbonic acid which attacks and dissolves rocks such as limestone. This is the process of carbonation.

$$CaCO_3 \quad + \quad H_2CO_3 \quad \rightarrow \quad Ca(HCO_3)_2$$
Calcium carbonate   Carbonic acid   Calcium bicarbonate

Carbonic acid combines with calcium carbonate to form calcium bicarbonate, which is soluble in water. The process of carbonation is reversible. Precipitation of calcite (a deposit also known as dripstone or **tufa**) happens in water saturated with calcium carbonate.

Oxidation

Oxidation occurs when minerals combine with oxygen. The process usually involves water in which oxygen is dissolved. Minerals may also take up oxygen directly from the air. Iron minerals are especially prone to oxidation. Oxidation destroys the structure of original minerals, weakening rocks and causing breakdown. It often attacks the iron-rich cements that bind sand grains together in sandstones.

Hydrolysis

Hydrolysis is a chemical reaction between rock minerals and water. Rock minerals known as silicates combine with water (rather than simply dissolving). The process is particularly active in the weathering of feldspar in granite. Secondary minerals or clays, such as kaolinite, illite and montmorillonite, are the products of the hydrolysis of silicates such as feldspar.

Hydration

Water molecules added to minerals form new minerals. For example, water added to anhydrite forms gypsum. The effect of hydration is to increase the volume of minerals. This sets up physical stresses in the rock. Hydration commonly causes surface flaking in rocks.

1.4 Biological weathering

Plants and animals play a key part in weathering. This **biological weathering** includes both physical and chemical processes.

- Physical processes include animals burrowing and tree root penetration. Animals may break up rock fragments by burrowing, and cause weathering by exposing fresh material to the atmosphere. More important are tree roots penetrating rock joints. As they grow and thicken, they prise rocks apart, causing block disintegration. In addition, when trees topple over, they exert leverage, which brings to the surface large masses of rock and soil.
- Dead organic material (particularly leaf litter) mixes with rainwater to form humic acids. These acids attack some rock minerals in a process called chelation. Where there is a dense vegetation cover (e.g. tropical rainforest), chelation is likely to be an important process contributing to rapid rates of weathering. Biological weathering even occurs in hot deserts.

2 *Mass movement*

2.1 Mass movement

Mass movement is the downhill transfer of slope materials as a coherent body. Some mass movements, such as landslides and mudflows, occur rapidly and are extremely hazardous. Others, such as soil creep and frost heave, occur slowly and pose little threat to people and property.

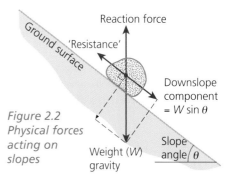

Figure 2.2 Physical forces acting on slopes

2.2 Causes of mass movement

Rapid mass movement indicates slope failure. To understand why this happens, we need to know that slopes represent a balance between **driving** and **resisting** forces. **Gravity** is the major driver, pushing

materials downslope. Its force varies with the sine of the slope angle (Figure 2.2). Other driving forces include the mass of material on the slope and water. Resisting forces act upslope. They include, for example, the strength of the slope materials, the frictional resistance between slope materials and the binding effects of vegetation.

Mass movement occurs when the downslope forces exceed the upslope forces. This triggers mass movement, which lowers the slope angle and restores stability. A number of factors may tip the balance and induce slope failure (Table 2.1). Increasingly, human factors are causal agents in mass movement events (Figure 2.3).

Table 2.1 Causes of mass movement and slope failure

Changes that increase the downslope force	Changes that reduce the upslope force
Steepening and undercutting at the base of a slope. This often results from natural erosion (e.g. a valley-side slope undercut by a river) and sometimes from human activity (e.g. a road cutting).	Heavy rain lubricates slope materials and reduces the friction between them and solid geology of the slope.
Loading a slope and thus increasing its mass. The common cause of loading is heavy rain which is absorbed by slope materials. Loading may also result from rockfall and construction on slopes.	Heavy rain increases water pressure within the pores of mineral materials, reducing their coherence and decreasing the resisting force.
	Earthquakes reduce the resistance of slope materials through violent shaking and often trigger landslides.

Deforestation removes the binding effect of tree roots on slopes and increases the amount of water absorbed by slope materials. Its effect is twofold: to reduce the resisting force and increase the driving force.

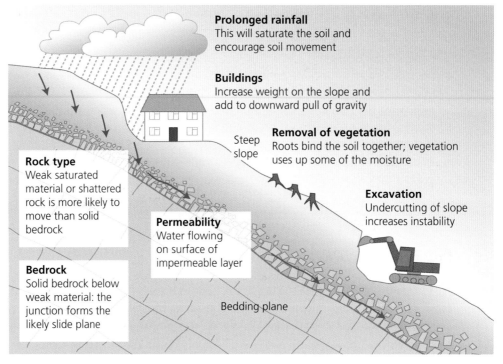

Figure 2.3 Human impacts on mass movement

2.3 Types of mass movement

Mass movements are classified according to their speed and moisture content (Figure 2.4). The most hazardous mass movements are slides and flows that move rapidly and give people little time to prepare for evacuation.

Slides and flows

Slides are masses of material that move across a clearly defined **slide plane**. This means that velocity is uniform throughout the sliding mass. **Flows** are mass movements that decrease in velocity with depth. Both slides and flows vary in their speed of movement, depending on their water content, particle size and the gradient of the slope.

Landslides usually occur in slope materials with low water content and across a curved slide plane. The moving mass often shows evidence of rotation with tilted black slopes, hence the description **rotational slide**. Rotational slides are common on steep slopes such as valley sides and coastal cliffs. Basal undercutting by either fluvial or marine erosion creates slope instability.

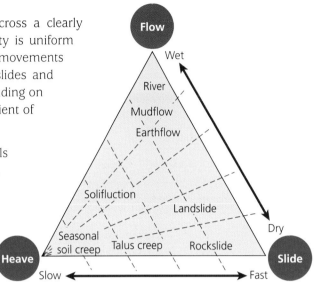

Figure 2.4 Types of mass movement

Mudflows are mobile, saturated flows of fine clay and silt. They are most common in areas of sparse vegetation cover. In hot deserts mudflows often form large fans at the foot of mountain slopes. In areas of active volcanism, where loose ash is easily transported by runoff and meltwater, mudflows are known as **lahars** and tend to follow river valleys.

Debris flows or **debris avalanches** comprise coarse regolith (including boulders more than 1 m in diameter) and other debris (e.g. timber). Rates of flow can reach 50 km/h. Debris flows are associated with extremely heavy rainfall events and rapid-run and, like lahars, are often confined to river valleys. They produce coarse, lobe-like deposits (alluvial fans) along mountain fronts.

2.4 Mass movement hazards

Small-scale landslides and other mass movement events are widespread. In the UK over 14,000 have been documented. Most of these are located in the uplands where slopes are steep and fluvial erosion is rapid, and in coastal areas where coastal slopes and cliffs are often oversteepened by marine erosion. Small-scale mass movements frequently occur after heavy rainfall. They damage property and block roads and rail lines, but are rarely a threat to human life.

In more extreme environments, mass movements are often major hazards and claim thousands of lives every year. Many mass movement hazards are the indirect or secondary result of primary hazards such as earthquakes, volcanic eruptions and hurricanes.

Table 2.2 Managing slopes to mitigate mass movement hazards

Scheme	Description
Re-vegetation	Slope stability can be improved by planting trees and other vegetation. Plant roots bind the regolith and increase resistance to gravity. Plants also intercept rainfall and transpire moisture, reducing the water content of slope materials.
Building retaining walls	Steep slopes (e.g. in road cuttings) may be stabilised by building retaining walls at the slope foot. The walls prevent undercutting and possible slope collapse.
Rock bolts	Rock bolts may be used to stabilise solid rock slopes.
Drainage	Improved drainage helps to reduce the loading of slopes by water after heavy rain.
Re-grading	Slope angles may be reduced to make slopes more stable. Terracing of slopes has a similar effect.

CASE STUDY 1   The Philippines, 2006

Physical details	Location: Ka'abag mountain, Guinsaugon, 670 km southeast of Manila in the Philippines. Date: 17 February 2006 The mudslide was 3 km wide and covered an area of 9 km². The slide was up to 30 m deep.
Cause	Heavy and continuous rainfall. Approximately 2 m of rain fell in 10 days. The cause was a La Niña event, with strong convection from the warm ocean in the western Pacific. Slopes were steep and unstable. Instability was caused by widespread deforestation over the previous 70 years. Deep-rooted trees had been removed and replaced by shallow-rooted coconut groves. Much illegal logging had occurred in the area.
Hazards	The primary hazard was a huge mudslide. The slide was fast-moving. There was no warning and local people did not have time to escape. Survivors described 'a wall of mud'. Half the mountainside collapsed in a few seconds.
Exposure	Exposure was high. Slopes around Guinsaugon are steep; rocks, which are deeply weathered, are susceptible to mass movement; rainfall events are intense. Population levels are high, with rural densities exceeding 200 persons/km². Recent population growth has been rapid, with annual rates of 2.7%. Expanding populations put pressure on resources: as a result, people are forced to live in areas exposed to mass movements.
Vulnerability	Warnings were issued by the Filipino government and many hill-side villages in the region were evacuated (20 people had been killed in the days immediately before the mudslide). Evacuation centres were set up after the disaster. They provided shelter, water supplies and health services. Emergency aid was provided by the Red Cross and Red Crescent. The remoteness of the disaster area hampered rescue attempts and increased vulnerability. Rapid population growth and overpopulation have encouraged deforestation and further increased vulnerability. In addition, widespread commercial (and illegal) logging had occurred in the previous decade. Although a logging ban had been established, it was not enforced and was undermined by corruption among politicians. In 2005, several villages threatened by mudslides had been evacuated. However, driven by poverty, people soon returned to their homes, despite the risks.
Impact	Approximately 1,000 people died and the village of Guinsaugon was destroyed. There were few survivors.

1 *Fluvial processes*

Rivers are energy systems. They derive their energy from:
- their volume of flow or discharge
- flow velocity

This energy is expended on the fluvial processes of erosion and transport. At low flow, rivers have only enough energy to overcome the frictional resistance of the channel and keep the water flowing. At high flow, rivers have surplus energy which is expended on sediment transport and erosion.

1.1 Fluvial erosion

Rivers erode, transport and deposit earth materials to produce distinctive landforms. Today, in temperate latitudes, rivers are the dominant landforming agents.

There are four principal fluvial erosional processes: abrasion (or corrasion), hydraulic action, cavitation, and solution (or corrosion).
- Abrasion mainly results from the transport of bedload. Like a giant grinding machine, the movement of cobbles and pebbles scours the channel bed and undermines the channel banks. These coarse particles are the 'tools' of erosion. Abrasion is a major cause of the bedload particles themselves becoming smaller and rounder (by attrition) as they move downstream.
- Hydraulic action describes the dragging force of flowing water, dislodging particles of sand and silt from the channel. Hydraulic action is effective where stream channels are formed from incoherent materials such as sand and gravel. It has little effect in channels cut in solid rock.
- Cavitation occurs when air bubbles implode in fissures and cracks in channel banks. The tiny shock waves they create weaken the banks and eventually lead to collapse.
- Solution is the chemical action of stream water, which dissolves carbonate rocks such as chalk and limestone.

1.2 Sediment transport

Load is the sediment transported by streams and rivers. We describe the size distribution of this sediment as the **calibre** of the load (Table 3.1).

Table 3.1 A river's load

Type of load	Sediments	Transport processes
Bedload	Coarse particles: boulders, cobbles and pebbles	Particles slide and roll along the channel bed at high discharge
Suspended load	Fine particles such as silt and clay	Particles are entrained (removed) at high discharge and in suspension in the flow
Solution load	Dissolved minerals from rocks such as chalk and Carboniferous limestone	Minerals are transported in solution. This type of transport does not depend on high river discharge

Sediment transport depends on flow velocity and particle size. A river's ability to erode and transport particles of a given size is known as its **competence**. Figure 3.1 summarises the relationship between flow velocity and particle size. A critical speed is required to remove and transport particles of a given size; this is the **erosion velocity**.
- Coarse particles (i.e. cobbles and pebbles) and fine particles (i.e. silt and clay) both require high erosion velocities.
- Sand-sized particles have relatively low erosion velocities.

Large particles such as cobbles and pebbles need high flow velocities to get them moving. Fine particles are also transported only at high velocities. This is because they stick together, bonded by tiny electrical charges. In contrast, sand is loose and incoherent and is eroded at low velocities.

The lower curve in Figure 3.1 shows the velocities needed for particles to stop moving or settle out of suspension.

- Coarse particles soon come to rest as flow speeds dip below the critical erosion velocity.
- Clay particles settle out of suspension only at very low velocities.

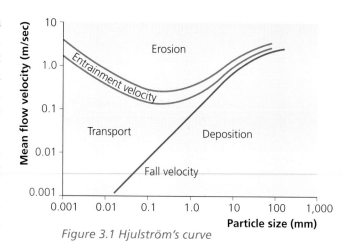

Figure 3.1 Hjulström's curve

The difference between erosion velocities and depositional velocities of particles has implications for sediment transport.

- Coarse particles will be in transit for relatively brief periods around peak discharge and are therefore likely to travel only short distances.
- Once entrained, fine particles will be transported long distances. The selective removal of fine particles from upland streams is one reason why boulders and cobbles dominate the load of upland streams.

Lakes and reservoirs are sediment traps. Therefore, rivers leaving lakes and reservoirs often contain no suspended load and bedload. As a result, these rivers have excessive energy that is expended through 'clear water' erosion.

2 *River channels*

2.1 Bankfull channel shape

River channels transfer water and sediment. Their shape in cross-section and plan are adjusted to carry the maximum discharge and sediment loads. **Bankfull discharge** determines the shape of river channel in cross-section. It is the maximum capacity of a river channel, reached just before the water overtops the banks and floods across the valley floor. On most British rivers, bankfull discharge occurs between 0.5 and two times a year.

Channel efficiency

- We calculate the **width/depth ratio** by dividing the width of the channel (from the top of each bank) by its average depth. Large values for width/depth ratio indicate relatively wide, shallow channels. These channels are inefficient for the transfer of water, but effective for transporting coarse bedload. Low width/depth ratios indicate more efficient channels, which are relatively narrow and deep.
- The **hydraulic radius** is the ratio of the bankfull cross-sectional area of a channel to its wetted perimeter. Wetted perimeter is the length of channel bed and banks in cross-section that are in contact with the water at bankfull. The higher the value of the hydraulic radius, the more efficient the channel for transporting water.

2.2 Cross-section channel shape and bankfull discharge

River channels respond to increases in discharge by adjusting their width, depth and velocity. These changes in width, depth and velocity are known as **hydraulic geometry**.

We can study the hydraulic geometry of a stream either:
- at a station (i.e. at one specific location), or
- at regular intervals downstream

As discharge rises, width, depth and velocity increase at different rates 'at a station'. Rates of increase depend on channel shape. In deep, narrow channels (often formed in coherent materials such as silt and clay), depth increases most rapidly. In wide, shallow channels (common where banks comprise incoherent gravel), the increase in width is more apparent.

Bankfull discharge also increases downstream and produces corresponding changes in width, depth and velocity. Generally, channel width increases more rapidly downstream than channel depth. Velocity also increases, despite the decrease in channel gradient. This is because river channels have:
- a more efficient shape downstream
- a smaller-calibre load

As a result, there is less frictional resistance to flow.

2.3 Pools and riffles

In alluvial channels there is an alternating sequence of shallows and deeps. The shallows that form high points in the stream bed are known as **riffles**. They consist of gravel and have relatively steep gradients. The deeps, with finer bed material, are **pools**. At low flow, the current is faster over the riffles than over the pools.

Pools and riffle sequences develop at high flow. Frictional resistance between the water and the channel bed and banks sets up turbulence, with alternating faster and slower flow. The spacing of these eddies is around six times channel width. The downstream changes in fast and slow flow lead to erosion in the pools (faster flow) and deposition in the riffles (slower flow).

Even in straight channels, water flows in a sinuous path, swinging from bank to bank. In the pool sections, where the flow is fastest, the sinuous path of the water causes bank erosion.

2.4 Channel planform

We have seen that a river adjusts to changes in discharge and sediment load by modifying its channel cross-section and channel gradient. A river also adjusts by changing its channel planform (Figure 3.2). There are three types of channel planform: straight, meandering and braided.

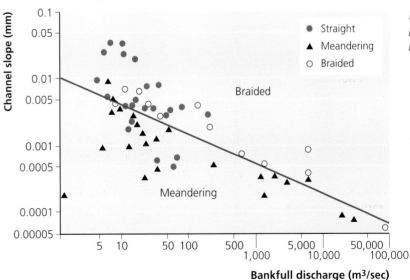

Figure 3.2 Channel planform and stream power

Channel planform is just one of five ways in which rivers can adjust their channels to accommodate increases in discharge and sediment load. Apart from planform, rivers can adjust:
- channel depth
- channel width
- cross-sectional shape
- channel gradient

Straight channels

Natural river channels are rarely straight for more than ten times their width. Straight channels occur when streams have low energy levels, small discharges and gentle gradients. There is minimal erosion, minimal sediment transport and little flow deflection.

Meandering channels

Meandering channels have a sinuous, wave-like form. Around 80% of river channels are classed as meandering. There are many theories of meander development. One of the most easily understood is that friction between flowing water and the channel bed and banks causes the **thalweg** (i.e. path of fastest current) to follow a sinuous path. Where the thalweg hits the channel bank, erosion occurs until eventually the channel takes on a sinuous form, as shown in Figure 3.3.

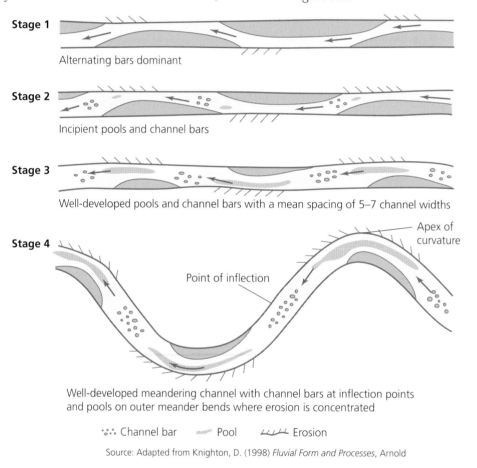

Source: Adapted from Knighton, D. (1998) *Fluvial Form and Processes*, Arnold

Figure 3.3 The development of meandering channels

Meandering channels most commonly occur where:

- bank material comprises coherent silts and clays (the more coherent the banks, the more sinuous the channel)
- channel gradient is moderately steep so that the river has power to erode its banks

Within meanders a secondary flow known as **helical flow** occurs, as shown in Figure 3.4. Helical flow is spiral-like motion subsidiary to the main downstream flow. A surface current moves across the river, elevating the water on the outside of the meander. This produces a return current close to the river bed, directed at the inside of the meander. The return current results in lateral accretion and the formation of a **point bar**.

Pools and riffles are features of meandering as well as straight channels. Pools occur on the outer bend of meanders, with riffles situated at the point of inflection. The spacing of pools and riffles in meanders adds weight to the theory that meanders develop from the sinuous flow found in straight channels.

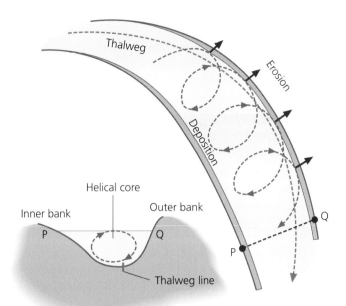

Figure 3.4 Helical flow in a meander

Braided channels

Braided channels consist of two or more channels divided by bars and islands, as shown in Figure 3.5. The cause of channel division is deposition of sediment within the channel. Braiding results from:

- an abundant bedload owing to poorly vegetated surfaces and coarse debris available from glacial erosion or volcanic eruptions
- easily eroded banks, especially banks comprising gravels and sand, which may cause a localised overloading of coarse sediment
- high and variable discharge, with high peak flows, often associated with meltwater in glaciated regions
- steep slopes

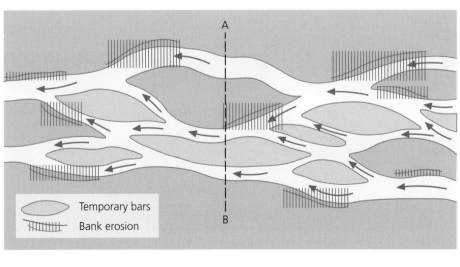

Figure 3.5 Braided channels

Braiding indicates that the river cannot transport its sediment load in a single channel. Deposition (**aggradation**) in the form of bars steepens the channel gradient and restores the river's competence to transport its bedload. Bars have a coarse upstream (proximal) end and grade to finer particles at the tail (distal end). They gradually migrate downstream.

Fluvial landforms

3.1 Fluvial landforms in upland regions

Geological structure and past erosional events (e.g. glaciation) have a strong influence on upland streams. The long profiles of upland streams are characteristically uneven. Waterfalls occur where weak rocks are eroded beneath a resistant caprock or where valleys have been overdeepened by glacial erosion. Rapids also occur where resistant rock bands cross stream channels.

The main fluvial landforms in upland areas are erosional. Upland valleys are often **V-shaped**, with steep valley slopes and narrow floors. Their distinctive shape results from rapid vertical incision — the result of steep gradients and high energy levels. **Backwasting** of valley slopes by mass movement and weathering helps to create an open, V-shaped valley cross-section. Where streams have meandering channels, vertical incision produces **interlocking spurs**.

3.2 Fluvial landforms in lowland regions

Floodplains

Most lowland rivers occupy broad valleys or **floodplains**, like the one in Figure 3.6. Floodplains result from both fluvial erosion and deposition. Lateral erosion and the downstream migration of meanders widen the valley, which is bordered by a line of steep slopes or **bluffs**. Fluvial deposition infills the valley with alluvium (silt, sand and gravel). This involves two processes: lateral accretion and vertical accretion.

- Lateral accretion mainly comprises point bars and channel sediments, abandoned as the river shifts its course through lateral erosion. These deposits are mainly coarse sands and gravels.
- Vertical accretion occurs when the water spills out of the channel and spreads across the valley floor. It consists of fine silts transported as suspended load.

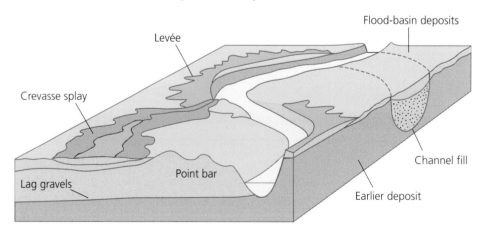

Figure 3.6 Floodplain deposits

Box 1 *Floodplain deposits*

Many floodplain deposits in northern Britain comprise two sediment units: a lower unit made up of layers of coarse gravel, and an upper unit of fine silt and clay. The lower unit represents old point bars and channel bars (bedload) abandoned as rivers migrate across the floodplain. The fine silts and clays at the top of the sequence are overbank sediments, i.e. the suspended load desposited in flood conditions.

Levées are low ridges of alluvium that run parallel to river channels. They form by vertical accretion. When water floods out of the channel at bankfull, it quickly loses energy. Coarse sediment is deposited nearest the channel and builds up to form levées. If floodwater surges through a levée, subsequent deposition may produce a fan-shaped area of alluvium known as a **crevasse splay**.

Channel fill makes a small contribution to floodplain development. Abandoned channels (e.g. cut-off meander loops or oxbows) are infilled with silt and plant remains.

Alluvial fans

An alluvial fan is a cone-shaped deposit usually found where a river emerges from a mountain course, as shown in Figure 3.7. In its mountainous course, a river may be confined by relief to a narrow channel. Once it emerges from the mountains, its valley widens, causing it to lose energy and deposit its load. A sudden change in gradient on leaving the mountains may also result in deposition. Alluvial fans are concave in profile. The coarseness of the sediment decreases with distance down the fan. Alluvial fans develop where there is abundant sediment supply. Where several valleys appear along a mountain front, alluvial fans may merge to form a continuous apron of debris known as a **bajada**.

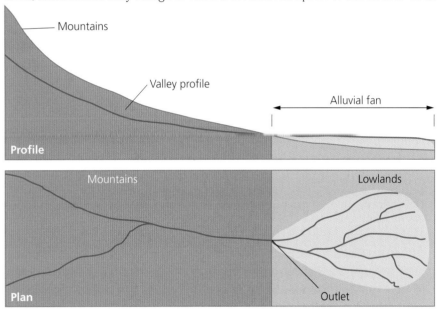

Figure 3.7 Alluvial fan development

3.3 Landforms caused by increased river energy

When rivers gain renewed energy, rates of erosion increase. Renewed energy may occur when rivers experience:

- an increase in discharge (e.g. climatic change producing higher rainfall, melting glaciers and icefields)
- a fall in base level (e.g. a lowering of sea level, tectonic uplift, isostatic uplift). This change is known as **rejuvenation**

Increases in discharge and rejuvenation create new erosional landforms. These landforms include incised meanders, river terraces and knickpoints.

Incised meanders

Although meandering is mainly associated with alluvial channels, **incised meanders** are found where a river cuts down into solid bedrock. There are two types of incised meander: intrenched and ingrown. **Intrenched meanders** occur where downcutting is rapid and little lateral erosion (or backwasting of valley slopes) occurs. **Ingrown meanders** develop where there is a slower rate of downcutting and where lateral erosion is significant. Tectonic uplift may accelerate the process of incision. Uplift of the Colorado Plateau in the last 12 million years is largely responsible for the deep incision of the River Colorado and its tributaries. This incision has produced the Grand Canyon and the classic intrenched meanders on tributaries such as the San Juan River.

River terraces

Renewed erosion may cause a river to incise into its floodplain. If this occurs rapidly, remnants of the original floodplain may be left as **river terraces** along the edges of the valley. When such terraces occur at the same level, they are known as **paired river terraces**.

Unpaired terraces form where lateral shifting of the river channel is rapid and is accompanied by some vertical erosion. Those parts of the former floodplain that survive form terraces, though not at the same level. Unpaired terraces in the British Isles often reflect climatic change and variations in river discharge.

Knickpoints

Rivers have a long profile that is graded to sea level (or to the confluence with larger rivers). If sea level falls or the principal river incises its channel, a river will adjust by cutting down to the new base level. This process starts at sea level (or at a confluence) and progresses upstream. Where the new graded profile intersects the old, an abrupt change of gradient develops. This feature is called a **knickpoint** (Figure 3.8).

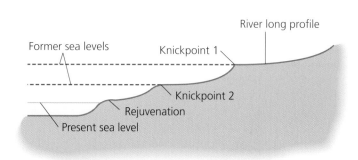

Figure 3.8 Knickpoints

Estuaries

Most river mouths in the British Isles form distinctive funnel-shaped **estuaries**. Estuaries are drowned lowland river valleys. Flooding occurred as sea level rose between 18,000 and 6,000 years ago, during the melting of Devensian ice sheets. In the last 6,000 years (a period of stable sea level), sedimentation has led to the growth of mudflats and salt marshes in estuaries.

Estuaries are sediment sinks and areas of low wave energy. The flood tide sweeps silt and clay into estuaries, which settles out at high water. Rivers carrying suspended sediment deposit fine particles as a result of **flocculation** (the mixing of fresh water and salt water, which causes the tiny silt and clay particles to stick together and fall out of suspension). A complex network of deep channels or creeks (rather like river channels) criss-cross the mudflats and salt marshes. They reach bankfull twice a day (at high tide) and discharge huge volumes of water.

Table 3.2 Causes of incision and aggradation

Incision	Aggradation
Less load to transport	**Too much sediment for river to transport**
Increase in vegetation cover	More erosion, weathering and mass movement on slopes
Decrease in mass movements	Mining spoil
Absence of inputs from wind	
No input of glacial drift	
Decrease in frost weathering	
Change in base level	**Change in base level**
Land rises (tectonics, isostatic change)	Land sinks
Sea level falls	Sea level rises
Increase in discharge and flow velocity	**Decrease in discharge and flow velocity**
Rainfall increases	Rainfall decreases
Urbanisation	Afforestation
Tectonic uplift	Dam construction
Deforestation	

Deltas

Along some coastlines, the volume of sediment deposited by rivers is too large to be removed by wave and tidal action. Under these conditions, a **delta** of river sediment builds out into the sea.

Deltas show great variation in planform (Figure 3.9). This reflects the importance of fluvial, wave and tidal controls. Deltas located on lakes and inland seas, where wave energy is low, are dominated by fluvial processes. In contrast, along coasts with high wave energy or a large tidal range, shoreline processes can disperse sediments deposited by rivers.

Deltas comprise two morphological elements: a delta front (the shoreline and the gently sloping offshore zone), and a delta plain that forms an extensive lowland made up of active and abandoned distributary channels. Delta front morphology is affected by the effectiveness of fluvial, tidal and wave processes, as shown in Figure 3.9.

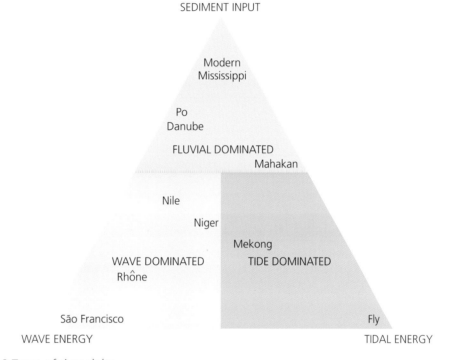

Figure 3.9 Types of river delta

The Mississippi is the only major delta almost solely determined in shape by sediment deposition by **distributary** channels. It has undergone very little modification by tidal and wave processes. Where wave action is more effective, a smooth arcuate shoreline is developed through the redistribution of river sediment by longshore currents (e.g. Nile Delta). Along coasts with strong tidal action, the delta front is dominated by ridges, channels and islands (e.g. tide-dominated Ganges-Brahmaputra Delta).

4   *River flooding*

Rivers flood when discharge exceeds bankfull channel capacity. The excess water spreads across the floodplain.

4.1 Causes of floods

A number of conditions are closely associated with river floods:
- high-intensity rainfall, creating rapid runoff
- prolonged rainfall on to already saturated soils
- snowmelt, where several weeks of accumulated precipitation is released into rivers

Some rivers are naturally more prone to flood than others. These **flashy** rivers, which have short lag times and high peak flows, have several common characteristics:

● upland catchments where mechanical (orographic) uplift intensifies convectional storms
● high drainage densities
● impermeable geology, e.g. boulder clay, gritstone, granite, with minimal groundwater storage
● steep slopes, which create rapid runoff
● a lack of vegetation cover (especially trees) with little interception, which would otherwise slow the movement of precipitation to rivers

During the summer months, intense thunderstorms in such drainage basins can create disastrous flash floods. These floods occur suddenly and without warning. Examples include the floods at Boscastle, Cornwall in August 2004 (Case Study 1) and in Ryedale, north Yorkshire in June 2005. Slow floods result from prolonged periods of rainfall or gradual snowmelt over several days. The severe floods on the River Severn in Gloucestershire and Worcestershire in July 2007 occurred after weeks of heavy rain in central Wales and the Midlands (Case Study 2).

Human activities can increase the frequency and severity of flooding. Deforestation, farmland drainage, the removal of wetlands, ploughing on steep slopes and urbanisation all increase the flood risk. In the past 50 years, uncontrolled building on floodplains has greatly increased the flood hazard in many parts of the UK.

4.2 Flood hazards

Floods are possibly the world's greatest natural hazard. Every year nearly 200 million people in more than 90 countries are exposed to severe floods — more than the number at risk from earthquakes and hurricanes. Between 1980 and 2000, floods caused an estimated 170,000 deaths.

CASE STUDY 1 Boscastle, Cornwall, 2004

Physical details	On 16 August 2004, peak discharge on the River Valency at Boscastle (Figure 3.10) reached 140 cumecs compared with an average flow of 0.5 cumecs. A 3 m wall of water swept down the river carrying trees, cars and other debris. This was a 1-in-400-year event. 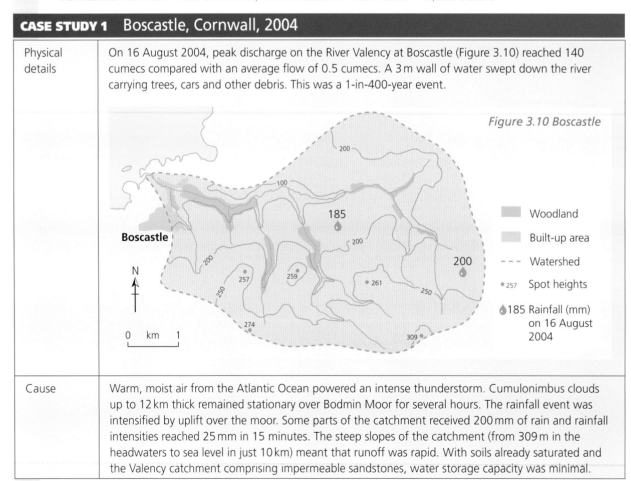 *Figure 3.10 Boscastle*
Cause	Warm, moist air from the Atlantic Ocean powered an intense thunderstorm. Cumulonimbus clouds up to 12 km thick remained stationary over Bodmin Moor for several hours. The rainfall event was intensified by uplift over the moor. Some parts of the catchment received 200 mm of rain and rainfall intensities reached 25 mm in 15 minutes. The steep slopes of the catchment (from 309 m in the headwaters to sea level in just 10 km) meant that runoff was rapid. With soils already saturated and the Valency catchment comprising impermeable sandstones, water storage capacity was minimal.

Hazards	Flash flood. Floodwaters and flood debris transported by the River Valency.
Exposure	Exposure was relatively high. A similar flood occurred further along the coast at Lynmouth in 1953 and killed 14 people. The Valency catchment covers the Bodmin upland, where slopes are steep, the rocks impermeable and there is little woodland cover. Mechanical uplift of warm, humid air masses as they cross the upland can transform thunderstorms into extreme rainfall events. The drainage basin is roughly circular, with the main tributaries of the Valency converging near the village. The population of Boscastle is approximately 2,000 in August, half of whom are residents and the rest tourists.
Vulnerability	The Environment Agency did not issue a flood warning. Violent but small-scale floods are difficult to forecast. Boscastle occupies the narrow floodplain, with many properties built adjacent to the river. During the flood a total of 150 people were airlifted to safety from stranded cars, rooftops and trees. The Environment Agency has taken mitigating action by constructing a larger flood relief channel on the Valency and will modify a narrow bridge where debris backed up the floodwaters.
Impact	Around 1,000 residents and visitors were affected. There was extensive damage to property and 25 businesses. Fifty-eight properties were flooded and four completely demolished. Four footbridges across the river were destroyed. There were no deaths or serious injuries. Telephone, gas and electricity supplies were disrupted. Around 50 vehicles were destroyed. There was long-term damage to tourism.

CASE STUDY 2   Lower Severn floods, Gloucestershire and Worcestershire, 2007

Physical details	The River Severn is the third largest British river, draining a catchment of nearly 11,500 km² (Figure 3.11). Mean annual rainfall exceeds 2,000 mm in the upper catchment. Large parts of the upper catchment in central Wales are above 300 m. Slopes are steep and the main rock types have low permeability. Runoff is rapid and small tributaries that feed into the Severn are highly responsive to rainfall and liable to flash floods. The flood wave peaked on 23 July, when Gloucester city centre narrowly avoided flooding.	*(map — Figure 3.11 The Severn catchment)* Shrewsbury · R. Severn · R. Severn Plynlimon 825 m · R. Teme · R. Avon Worcester · Evesham · Tewkesbury · Gloucester 0 km 50 · Severn estuary

Figure 3.11 The Severn catchment

Cause	The floods followed an extreme rainfall event on 20 July, when 135 mm of rain fell at Pershore (near the confluence of the Avon and Severn) in just 16 hours. This extreme rainfall event had a recurrence probability of only 0.1% a year. Flash floods in several small catchments in Worcestershire fed into the major rivers, raising the Severn at Worcester nearly 6 m above normal and the River Avon at Evesham to the highest level ever recorded. The severity of flooding was due to previous heavy rainfall (2007 was the wettest summer since 1766): soils had already been saturated by heavy rains in June. Where flood defences were overtopped, it was because the flow level exceeded their design.
Hazard	River flood.
Exposure	The lower catchment, especially between Worcester and Gloucester, with major tributaries such as the Avon and the Teme, is one of the most flood-prone areas in the UK. There is an extensive floodplain in this area which is only a few metres above sea level. Significant urban development on the floodplain, especially at Tewkesbury and Upton-upon-Severn, increases exposure.

Vulnerability	The flood risk has been increased by building on the Severn floodplain. Recent urban growth has accelerated the loss of floodplain land, reducing natural water storage and placing more properties at risk. In the lower Severn and Avon catchments, changes in rural land use have also increased vulnerability: improved drainage of farmland quickly transfers rainwater into the rivers; the conversion of pasture to arable land has increased runoff; and clear felling of forestry plantations has had a similar effect.
	The lack of hard flood defences also makes the lower Severn valley vulnerable to flooding. There are few flood embankments, and other forms of river engineering such as flood relief channels, channel widening and straightening are absent. Flood mitigation is both long and short term.
	Immediate responses to the flood event included the erection of temporary flood barriers and flood warnings issued by television, radio, the Environment Agency's Floodline and the internet. After the floods, Severn Trent water authority distributed 5 million litres of bottled water a day via 1,500 water bowsers. The government provided £87 million of emergency aid for funding schools, transport and businesses hardest hit by the floods. The EU contributed a further £31 million in compensation.
Impact	By 21 July flooding had occurred in towns and villages on the Avon's floodplain to a depth of 2 m. At Tewksbury, floodwaters entered the town's medieval abbey for the first time in 247 years, and there was widespread flooding all along the River Severn from Upton-upon-Severn to Gloucester. Flooding of an electricity substation near Gloucester left 50,000 households without power (some for up to 2 days). A water treatment plant in Tewkesbury was also flooded, with the result that water supplies for 140,000 households were cut off and without water for at least 5 days. Some 10,000 motorists were left stranded on the M5 and surrounding roads and forced to abandon their cars due to floodwater and landslides. The total insured losses were estimated at between £1 and £1.5 billion.
	There was large-scale damage to property and disruption of businesses. About 27,000 domestic insurance claims and 6,800 business claims were made. Crops were submerged and maize, potatoes and hay crops lost.

4.3 Flood management

There are two approaches to flood management: flood abatement and flood protection.

Flood abatement

Flood abatement aims to slow the movement of water into river channels. Its effect is to lengthen lag times and reduce peak discharge. This can be achieved through land use change and land use management in the upper catchment.

- Afforestation increases interception, evaporation and transpiration, reducing runoff and lengthening lag times.
- Changes in agriculture, from arable to pasture, have a similar effect.
- Land use management in farming, such as terracing and contour ploughing, also helps to reduce the risk of flooding.

Flood abatement policies often run into problems where land is privately owned or in the hands of several owners. In these circumstances, it may be difficult to get agreement on changes in land use and farming practices. Whole-catchment planning may also be hindered where administrative and political boundaries do not coincide with the watersheds of river basins. Flood abatement is a long-term strategy; in the short term, it may do little to reduce flooding.

Flood protection

Flood protection includes structural and non-structural approaches.

Structural approaches

Structural approaches involve hard engineering (Table 3.3 and Figure 3.12), which aims to confine floodwater to river channels or divert it to temporary storage in reservoirs and on floodplains. Structural approaches are expensive to implement and may have undesirable environmental effects. However, some, such as dam building, may offer a range of other benefits such as providing water supplies, opportunities for recreation and leisure and the generation of hydroelectric power.

Table 3.3 Structural approaches to flood protection

Scheme	Description
Embankments/levées	Embankments on either side of the channel to increase channel capacity. Potentially hazardous if the water level is above the floodplain. Unless set back from the river channel, levées will raise flood levels and increase the flood risk.
Channel straightening/ channelisation	Removing meanders steepens the average gradient, increases the flow velocity and scours and deepens the channel. Eventually, meanders will re-form unless the new channel is lined with concrete.
Flood relief channels	Artificial channels that take some of the floodwater relieve natural channels, e.g. Jubilee River on the Thames in Berkshire.
Sluice gates	Sluice gates are raised during times of flood to protect settlements downstream. Water is diverted into flood basins or washlands, where it is stored temporarily, e.g. River Wyre at Garstang in Lancashire.
Dam building	Storage of floodwaters in reservoirs or flood storage basins. Dams and reservoirs offer multipurpose usage, such as water supply, hydroelectric power, and recreation and leisure, as well as flood prevention (e.g. Kielder Water in Northumberland).
Channel enlargement	Increasing the width and depth of river channels through dredging to provide greater capacity. 'Clearing and snagging' are also used to remove vegetation and other debris from river channels, which increases channel cross-sectional area and flow velocity.

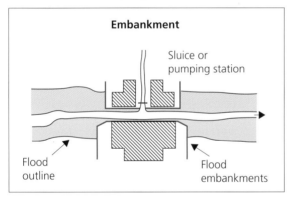

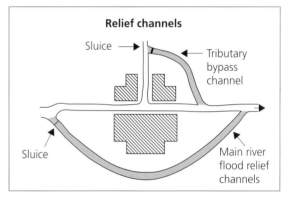

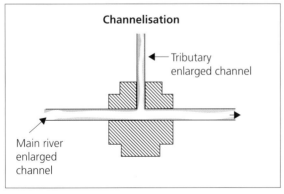

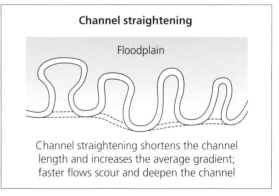

Figure 3.12 Hard engineering schemes

Non-structural approaches

Non-structural approaches to flood protection are increasingly favoured. They are cheaper, less environmentally damaging and more sustainable than hard engineering alternatives. The main non-structural approaches are summarised in Table 3.4.

Table 3.4 Non-structural approaches to flood protection

Scheme	Description
Floodplain management	Floodplains are attractive for settlement because they offer flat land that is easily developed and are locations for river crossings and transport routes. Pressure to develop floodplains in the UK, driven by population growth, societal changes to family structure and land shortages, is strong. However, floodplain development is exposed to significant flood risks, which will increase with climate change. Moreover, floodplain development drains the wetlands that are natural storage areas for floodwater. Increasingly, planners in the UK guide new development away from floodplains, arguing that in the long term such development is unsustainable.
Flood forecasts and flood warnings	In England and Wales, the Environment Agency operates a flood warning system. Flood warnings are issued by the agency for rivers through its telephone Floodline and its website. There are three levels of warning: flood alert, flood warning and severe flood warning. The system encourages people and organisations at risk to take action (e.g. evacuation) to mitigate the worst effects of flooding. Maps published online by the Environment Agency show the flood hazard areas and allow households to assess the general level of flood risk to their property.
Flood insurance	One mitigating option for people and organisations exposed to flood risk is flood insurance. In the UK, insurance companies are using GIS to provide more accurate estimates of the flood risk to individual properties. Typical flood insurance covers damage to property, loss of life, debris removal, relocation expenses and the cost of materials and equipment such as sand bags and pumps.

TOPIC 4 Cold environments

1 Distribution of cold environments

Cold climates that support glaciers, ice sheets and permafrost occur in high-latitude and high-altitude areas. The largest expanses of snow and ice are in the polar regions, principally in the ice sheets of Antarctica and Greenland. Most of the world's highest mountain ranges — Himalayas, Andes, Alps and Rockies — also support icefields and glaciers. Periglacial environments occupy sub-Arctic Canada, Alaska and Siberia. In these regions glaciers and ice sheets are absent, but temperatures are so low that the ground is permanently frozen.

2 The chronology of glaciation

For 90% of the last 1.6 million years, middle and high latitudes have experienced glacial conditions. During this period (known as the Pleistocene), there have been at least four major **glacials** or ice ages and three warmer spells or **interglacials**. In addition, there have been many cold and warm phases lasting just a few thousand years. These shorter episodes of cold and warmth are called **stadials** and **interstadials** respectively.

The most recent glacial (the Devensian) lasted from approximately 120,000 to 10,000 years ago. Following **deglaciation**, the world entered a warm interglacial phase (the Holocene) that has continued to the present day. Figure 4.1 shows the patterns of ice sheet movement and glacial erosion in Britain.

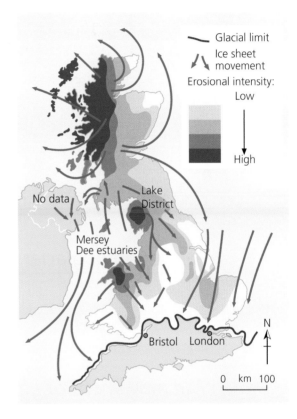

Figure 4.1 Patterns of ice sheet movement and glacial erosion in Britain

Box 1 The causes of ice ages

A combination of four factors causes ice ages.

- Changes in the output of solar radiation. Glaciations seem to occur when sunspot activity reaches a maximum.
- Changes in the **albedo** (reflectivity of the Earth's surface). The accumulation of snow and ice increases reflectivity and promotes cooling of the atmosphere. There is a positive feedback effect: the snow and ice cover lowers temperatures, which increases snowfall and amplifies the initial change.
- Changes in the Earth's orbit and the tilt of the Earth's axis. Over a period of nearly 100,000 years, the Earth's orbit varies from circular to elliptical. Glaciations occur when the orbit is more elliptical. The tilt of the Earth's axis varies from 22 to 25° over a period of 42,000 years. A more extreme tilt gives greater seasonality, strengthening the general circulation and decreasing the likelihood of glacials. Finally, variations in the orientation of the Earth's axis (comparable to the wobble of a spinning top) occur in a cycle of 21,000 years. This determines which hemisphere is closer to the sun in winter. When the northern hemisphere, with its larger landmasses, is nearer the sun in winter (the situation today), glacials are less likely to occur.
- Changes in the circulation of surface ocean currents, e.g. the breakdown of the warm North Atlantic conveyor.

3 Types of glacier

At the glacial maximum 20,000 ago, most of northern Europe was submerged under a vast sheet of ice. This was **continental glaciation**, with only the peaks of the highest mountains (known as **nunataks**) protruding above the **ice sheet**. Today, continental-type glaciation is confined to Antarctica and Greenland. The Greenland ice sheet reaches a maximum thickness of 2.5 km.

In plateau-like regions such as the Vatnajökull in Iceland and the Columbia plateau in Alberta, small ice caps or **icefields** develop. Glaciers nourished by these icefields spill over the edge of the plateaux as **outlet glaciers**. Sometimes outlet glaciers merge in the lowlands to form **piedmont glaciers** (Figure 4.2).

In steep mountain ranges such as the Himalayas and Alps, glaciers occupy only the valleys and the flanks of the mountains. Ice accumulates in sheltered circular hollows as **cirque glaciers**. If the mass of ice is sufficient, these glaciers will extend downslope to form

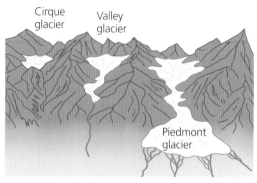

Figure 4.2 Different types of glaciers

valley glaciers. Such spatially limited glaciation is known as **Alpine glaciation**. **Niche glaciers** are small glaciers that cling to gullies and hollows on mountain faces. In the British Isles the breakdown of the North Atlantic conveyor (warm ocean current) led to a brief but intense cold snap known as the **Loch Lomond stadial** between 11,000 and 10,000 years ago. Small valley glaciers and cirque glaciers re-formed in the Scottish Highlands, Lake District and Snowdonia.

4 Formation of glacier ice

Glacier ice forms when the annual accumulation of snow exceeds melting. Under the weight of successive snowfalls and over many years, low-density snow (density less than 0.1) is compacted. First it forms granular snow (density around 0.3), then **firn** or **neve** (density 0.5) and finally glacier ice (density 0.9). Unlike snow, glacier ice can flow downhill under its own mass and gravity.

5 The movement of glaciers

Glaciers move in two ways: by internal deformation and by sliding.
- In climates of extreme cold, glaciers are frozen permanently to the underlying bedrock, meaning these cold-based glaciers are incapable of much erosion. This explains how delicate landforms such as tors can survive glaciation. Cold-based glaciers move by internal deformation caused by the weight of ice and gravity: ice crystals in the glacier rearrange into parallel layers and slide past each other.
- In warm-based glaciers, 90% of movement occurs through basal sliding. A film of water at the base of the glacier (Box 2) lubricates the ice and causes sliding. Therefore, glaciers in temperate regions can move quite rapidly. For example, the Nisqually Glacier on Mt Rainier in Washington state moves up to 50 cm a day. Slope steepness and ice volume also influence rates of glacier flow. Warm-based glaciers are active agents of erosion.

Box 2   *Pressure melting*

Pressure causes ice to melt at temperatures below zero. This explains why water is present at sub-zero temperatures at the base of many glaciers. In temperate glaciers, ice in contact with the bed often undergoes pressure melting. This allows the glacier to slide on its bed. Constrictions in glacier flow also increase pressure and cause melting. Equally, any reduction in pressure will lead to freezing. Pressure melting and refreezing, which is common in warm-based glaciers, is known as **regelation**.

Mass balance

Mass balance is the difference between a glacier's annual accumulation and ablation (melting, subli-mation) of snow. There are three states of mass balance:

- positive: accumulation exceeds ablation, increasing the ice mass and causing the glacier to advance
- negative: ablation is greater than accumulation; the glacier shrinks and retreats upvalley
- neutral: accumulation and ablation are equal, causing the glacier's terminus to remain static

The upper part of a glacier, where ice accumulation exceeds ablation, is the **accumulation zone** (Figure 4.3). In the lower part or **ablation zone**, ablation is greater than accumulation. The boundary between the accumulation zone and the ablation zone is the equilibrium or **firn line**. Above the firn line, the glacier surface is snow-covered.

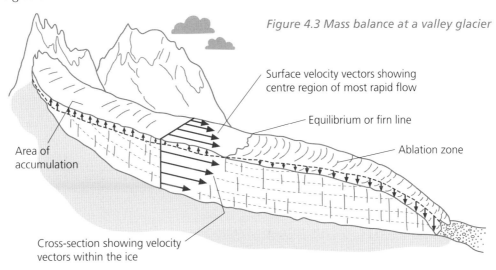

Figure 4.3 Mass balance at a valley glacier

Surface velocity vectors showing centre region of most rapid flow

Equilibrium or firn line

Ablation zone

Area of accumulation

Cross-section showing velocity vectors within the ice

Box 3   *Glacier advance and retreat*

An increase in accumulation consequent on a fall in temperature or an increase in precipitation results in glacier expansion and advance. Equally, glaciers will shrink and retreat if there is an increase in temperature or a fall in precipitation. Currently, most valley glaciers are retreating in response to global warming. However, at a local scale individual glaciers may respond to heavy winter snowfall by advancing. The Nisqually Glacier on Mt Rainier retreated steadily from 1840 to 1951. However, following high snow accumulation between 1944 and 1951, the glacier advanced to its present position, where it has remained more or less stationary for the last 20 years.

Glacial erosion

7.1 Processes of glacial erosion

Glaciers erode in two ways: by abrasion and by quarrying.

- Rock particles frozen into a glacier and particles dragged along at the base of the glacier scour and abrade the bedrock. This process of abrasion rounds and smoothes rock outcrops. Fine-grained particles may polish the bedrock and coarser particles may leave deep scratches called **striations** or **striae**. Striae provide important clues about the direction of ice flow.
- Quarrying (or plucking) removes bedrock particles along joints and bedding planes. For example, a small rock outcrop at the base of a glacier will cause frictional resistance to ice flow. On the upstream side, high pressure results in melting and the meltwater runs into joints and fissures in the rock. On the downstream side where pressure is lower, refreezing occurs. Therefore, as the glacier moves forward rock is quarried along joints in low pressure areas.

Roches moutonnées provide evidence of quarrying. They are small rocky outcrops that have been smoothed and steepened by glacial erosion. The upstream side (**stoss**), exposed to intense pressure, is abraded and smoothed, as shown in Figure 4.4. The downstream side is steep and uneven due to the effects of quarrying. The process of pressure melting and refreezing is known as **regelation**.

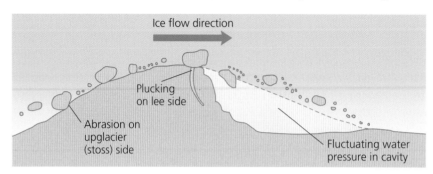

Figure 4.4 The location of abrasion and quarrying (plucking) on the lee side of a roche moutonnée

7.2 Landforms of glacial erosion

Cirques

Cirques (also known as corries, coires and cwms) are deep, amphitheatre-like rock basins cut into mountainsides. Formed by glacial erosion, cirques in the northern hemisphere occur most frequently on north- and east-facing slopes.

Cirque glaciers developed on these slopes because of:
● the accumulation of blown snow from prevailing southwesterly winds
● the colder microclimate (e.g. longer periods of shadow)

A possible sequence of events leading to cirque formation involves:
● freeze–thaw beneath a snow patch (nivation) and the removal of debris by surface wash and solifluction to create a shallow depression
● snow turns to firn and glacier ice and slides/flows downslope, overdeepening the depression by abrasion and quarrying
● the backwall of the overdeepened depression retreats by freeze–thaw weathering, while glacial erosion further deepens the depression (Figure 4.5)

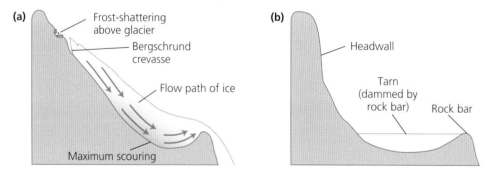

Figure 4.5 Cross-section of a cirque glacier (a) During the early stages of glaciation, the hollow becomes progressively overdeepened. (b) After glaciation, the overdeepened basin is occupied by a lake or tarn

In a fully developed cirque glacier, the weight of ice in the upper section of the glacier causes a rotational sliding movement that overdeepens the rock basin by abrasion. However, at the outlet of the basin, where ice movement is directed upwards, erosion is less severe and a rock bar or **lip** forms. A deep crevasse known as a **bergschrund** develops between the glacier and the headwall. Meltwater accumulates here and may assist quarrying on the headwall.

If two adjacent cirque glaciers cut back their headwalls and converge, they may reduce a broad ridge to a knife-edged feature called an **arête** (e.g. Crib Goch in Snowdonia). When three or more cirque glaciers converge in this way, they form a **pyramidal peak** (e.g. the Matterhorn).

Glacial troughs

Glacial troughs (or glacial valleys) are channel landforms, carved through solid rock by valley glaciers and ice streams (areas of more rapid flow within ice sheets). They probably developed along valleys that existed prior to glaciation. Glacial troughs are:

- roughly parabolic or U-shaped in cross-section — the result of glacial erosion of both the valley sides and the valley floor
- deeper than tributary valleys occupied by smaller glaciers. Because erosion rates of glaciers in major valleys were greater than those of smaller tributary glaciers, on deglaciation these tributary valleys are left hanging above the main glacial trough
- straighter than river valleys, with projecting spurs planed off or truncated
- irregular in long profile. The upper part of the trough often has a steep rock wall known as a trough head. Trough heads form where glacier ice from extensive icefields converges, increasing its speed and power of erosion. Further downvalley, rock basins are carved in areas of more intense erosion (e.g. where the valley narrows or tributary glaciers join the main glacier). Following deglaciation, these basins may form ribbon lakes

Diffluent cols

In some circumstances, valley glaciers erode breaches through watersheds. If a valley is blocked by ice or a constriction restricts ice flow, a valley glacier may overflow and breach the lowest point on a watershed. This process, known as **glacial diffluence**, forms a col or narrow gap in a watershed. The Lairig Ghru is such a col. It forms a spectacular 400 m breach across the main watershed of the Cairngorms.

Ice sheets and glacially eroded landscapes

Erosion by ice sheets often produces landscapes very different from those eroded by valley glaciers and cirque glaciers. Extensive areas such as the Laurentian Shield in Canada and the Baltic Shield in Scandinavia show the effect of abrasion and quarrying by ice sheets. Typical of these glacially eroded landscapes are streamlined roches moutonnées and interspersed rock basins occupied by thousands of small lakes. Their orientation often follows the direction of ice flow for hundreds of kilometres. Such a landscape of rock ridges and shallow lakes is called **knock and lochan topography**.

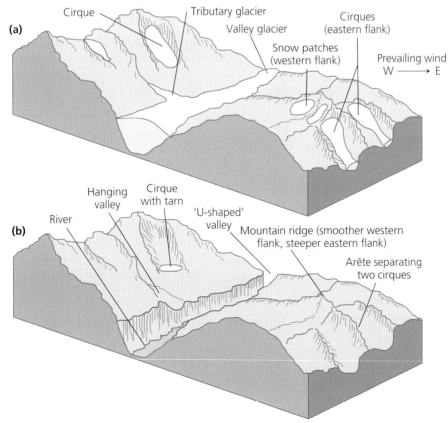

Figure 4.6 Landforms developed by Alpine glaciation: (a) during glaciation and (b) after deglaciation

Glacial deposition

Glaciers are like conveyor belts for transporting rock debris. This rock debris derives from:

- glacial erosion of the valley sides and valley floor
- rockfall due to weathering of valley slopes
- rock avalanches and other mass movements from valley slopes

Eventually, this rock debris is deposited:

- at the ice front or terminus of the glacier
- under the glacier
- under the ice and beyond the ice front by meltwater flowing within and from the glacier

Collectively, these deposits are called **glacial drift**. Large particles of rock transported by ice, and which are of different geology to the area in which they are deposited, are known as **erratics**. Erratics provide valuable evidence about the flow of ice during glacial periods. For example, we know from the distribution of erratics of Shap granite (Cumbria) that ice from the Lake District flowed south into Lancashire and Cheshire, as well as east across the Pennines into the Tees Lowlands.

8.1 Ice-contact depositional landforms

Rock debris deposited in contact with the ice comprises an unsorted mix of particles of all sizes. This material is known as **till** or **boulder clay**. It may be supraglacial, englacial or subglacial. Some till is also bulldozed downvalley ahead of the glacier. **Moraine** is till that has been piled into a variety of hummocky mounds and ridges.

Box 4   *Types of rock debris transported by glaciers*

- Supraglacial: material carried on the surface of the glacier.
- Englacial: material carried within the glacier. This debris may enter the glacier through crevasses or melt its way into the glacier.
- Subglacial: material carried at the base of the glacier and largely responsible for abrasion.

Lateral moraines

Lateral moraines form along the sides of valley glaciers. They consist of piles of loose rock debris derived from rockfall and avalanching on to the glacier from the adjacent valley slopes. If the glacier recedes or shrinks, this debris forms prominent ridges that run parallel to the valley side. The lateral moraines near the terminus of the Athabasca Glacier in Alberta stand over 120 m high and are 1.5 km long.

Terminal moraines

Moraine is delivered to the **snout** of a valley glacier by the forward movement of glacial ice. If the snout remains stationary for prolonged periods, the moraine accumulates to form a low ridge or **terminal moraine** across the valley. Terminal moraines comprise the melt-out of debris-filled ice and debris pushed by an advancing glacier. Glaciers retreating intermittently leave behind a succession of smaller **recessional moraines**.

Medial moraines

Where two glaciers meet, two adjacent lateral moraines merge to form a central **medial moraine** on the glacier surface.

Hummocky moraines

Hummocky moraines form chaotic landscapes, often made up of hundreds of steep-sided mounds (up to 50 m in height). They have no consistent orientation or linear development. These moraines are associated with dead or wasting ice. The largest area of hummocky moraine in Britain is in Glen Torridon in northwest Scotland.

Drumlins

Drumlins are smooth, oval-shaped hills made of till. They have a streamlined form elongated in the direction of ice flow. Typical dimensions are 5–50 m in height and 1–2 km in length. In profile, drumlins have a short, steep slope (stoss) which face up-glacier and a gentle long slope (tail) which face down-glacier. The typical shape is shown in Figure 4.7. Drumlins occur in swarms often in lowlands (e.g. Vale of Eden) close to centres of ice dispersal.

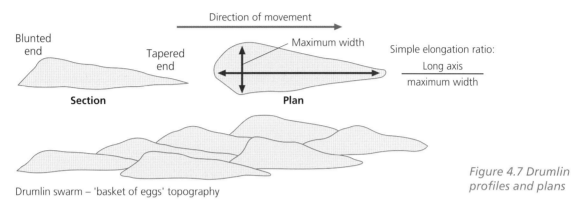

Drumlin swarm – 'basket of eggs' topography

Figure 4.7 Drumlin profiles and plans

There are several theories of drumlin formation. The most widely accepted suggests that moraine beneath an ice sheet is eroded into streamlined forms by moving ice.

8.2 Glacio-fluvial depositional landforms

Glacio-fluvial deposits are laid down by meltwater and sorted by particle size into layers. Meltwater is an important agent of sediment transport and deposition in warm-based glaciers and during phases of deglaciation. Meltwater flows on the surface of glaciers (**supraglacial**), in tunnels within glaciers (**englacial**), at the base of glaciers (**subglacial**) and beyond the ice front (**proglacial**). Landforms that result from glacio-fluvial deposition fall into two groups.

- Pro-glacial features: deposited by meltwater beyond the ice front (e.g. sandur and valley sandur).
- Ice contact features: deposited by meltwater within glaciers (e.g. eskers, kames and kame terraces). These are shown in Figure 4.8.

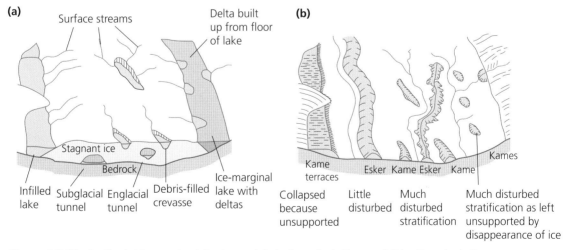

Figure 4.8 Glacio-fluvial ice contact features: (a) during glaciation and (b) after glaciation

Sandur

Sandur (or outwash plains) are extensive spreads of sand and gravel laid down by meltwater streams beyond the margin of glaciers and ice sheets. In southern Iceland, powerful meltwater streams draining several small icefields have deposited huge amounts of coarse sediment along the coast. As a result, the coastal plain has been extended several kilometres seaward. Proglacial lakes impounded between the ice front and recessional moraines also encourage rapid sedimentation by meltwater streams. Valley sandur develop in steep-sided glacial troughs that limit their lateral expansion.

Eskers

Eskers are sinuous ridges of sand and gravel that are often several kilometres long. They derive from sediments deposited within the channels of meltwater streams that flowed on, through or at the base of glaciers. Some eskers appear to run upslope — clear evidence that meltwater, under hydrostatic pressure, can flow uphill.

Kames and kame terraces

Cavities such as crevasses within glaciers may fill with sediment. Subsequent melting of the glacier causes the sediment to collapse to form an isolated mound or **kame**. Where sediments accumulate along ice margins (e.g. in a marginal lake between a glacier and the valley side), the eventual retreat of the ice leaves a continuous embankment parallel to the valley side. This is a **kame terrace**. The melting of ice cores trapped beneath glacial deposits may create surface depressions or **kettle holes** that later form small lakes.

8.3 Meltwater erosional landforms

Meltwater (or overflow) **channels** are the main glacio-fluvial erosional landform. These channels are subglacial in origin and were cut by meltwater streams with very high discharge. Meltwater channels are usually steep-sided, deep and fairly straight. In long profile, some have up-gradients. This shows the influence of hydrostatic pressure, enabling water to flow uphill if there is enough force behind it. The high discharges needed to form large meltwater channels may be related to the formation of temporary lakes beneath glaciers. The North York Moors has several overflow channels cut by meltwater at the end of the last glacial period. The largest, Newtondale, is 15 km long, 80 m deep and was formed by meltwater draining pro-glacial Lake Eskdale. At its peak, discharge reached 10,000 cumecs.

9 *Periglaciation*

In high latitude and some high-altitude regions not covered by ice, temperatures are so low that the ground is permanently frozen. In this **periglacial** environment, processes of freeze–thaw and the growth of ice masses in the ground give rise to distinctive landforms.

Figure 4.9 shows that the most extensive periglacial areas today are in northern Canada and northern Siberia. However, periglacial areas were far more widespread in the past. For example, southern Britain, although ice-free, had an intensely cold climate throughout the last glacial period. Most periglacial landforms in southern Britain are relict features from this period.

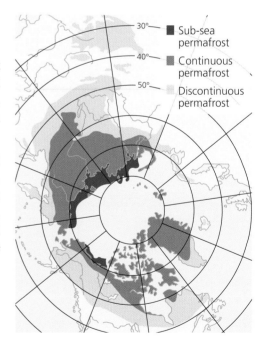

Figure 4.9 The distribution of the main permafrost types in the northern hemisphere

9.1 Periglacial processes

Permafrost

Permafrost is perennially frozen ground. Regions of continuous permafrost usually have a mean annual temperature of −5°C and below. Permafrost consists of two important layers: the **active zone** and the **frost zone**.

- The active zone lies near the surface and above the frost zone. It is in this zone that freeze–thaw occurs. The frost melts during the summer before refreezing in the autumn.

- The frost table separates the active layer from the permanently frozen layer or frost zone. Unfrozen areas within the frost zone are known as **taliks**.

Ground ice
In periglacial areas, ice often exists as large segregated masses in the ground. When water freezes, its volume increases by 9%. If the ground has a high moisture content, the growth of ice crystals will attract remaining liquid water and lead to the development of segregated ice as lenses and veins. The result is local expansion (freezing) and contraction (melting) of the surface — a process that can produce a number of periglacial landforms.

Frost weathering
Frost weathering is an important process in periglacial environments. Numerous freeze–thaw cycles and the sparse cover of soil and vegetation mean that frost action is very active. Landforms caused by frost weathering include **screes** (talus) and **blockfields**. In upland Britain, most screes and blockfields are relict features formed during cold conditions in the late glacial period.

Frost cracking
Sub-zero temperatures cause the ground to crack by contraction. This process produces polygonal cracks similar to those formed in drying mud. Frost cracking is a major cause of patterned ground.

Mass movement
Solifluction is one of the most effective processes in periglacial environments. It is defined as 'the slow flowing from higher to lower ground of masses of waste saturated with water'. Solifluction operates on slopes as gentle as 1 or 2° and on fine sand and silt material. Movement is confined to the active layer. Rates of flow vary with climate, slope and vegetation cover, but are usually of the order of 1–10 cm a year. In active periglacial environments, frost creep (heave) and permafrost (which creates an impermeable zone, causing saturation of the active layer in summer) combine with solifluction. This modified mass movement process is known as **gelifluction**.

9.2 Periglacial landforms
Patterned ground
Patterned ground, shown in Figure 4.10, is a characteristic feature of periglacial environments. It describes the distribution of rock particles in systems of polygons, nets, steps, stripes and circles. Each of these features is sub-divided into sorted and unsorted forms.

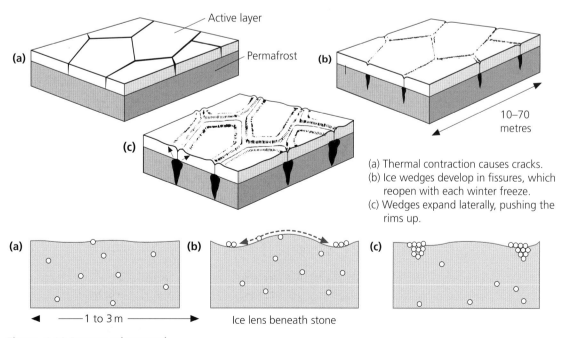

(a) Thermal contraction causes cracks.
(b) Ice wedges develop in fissures, which reopen with each winter freeze.
(c) Wedges expand laterally, pushing the rims up.

Figure 4.10 Patterned ground

Circles, nets and polygons normally occur on flat surfaces; steps and stripes form on slopes of between 5 and 30°. Frost cracking and frost heave are important processes in the development of patterned ground. Frost heave pushes larger stones to the surface and, because of the cambering of the surface, stones then move laterally. On steeply sloping ground, this cambering is oriented downslope. Coarser particles raised to the surface by frost heave roll into the depressions between the cambers. The result is alternating stripes of coarse and fine particles.

Ground ice phenomena

Ice wedges are downward-tapering bodies of ice up to 10 m in depth. In plan they form a polygonal pattern at the surface. They appear to result from frost cracking. When an ice wedge melts, it may fill with sediment to form an **ice wedge cast**.

The formation of ice lenses in the active layer can heave the overlying sediments into small symmetrical mounds (3–70 m high) known as **pingos**. When the ice lens eventually melts, it leaves a circular depression or **ognip**. This process is shown in Figure 4.11.

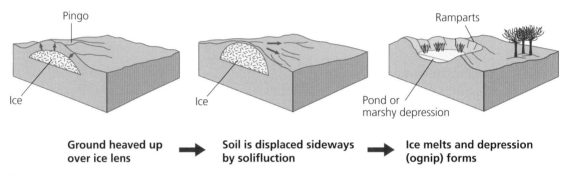

Figure 4.11 Formation of pingos and ognips

Two theories describe the formation of pingos:
- Closed-system pingos develop beneath former lakes. Initially, the water in the lake prevents the surrounding regolith from freezing. Eventually, the lake fills with sediment. This reduces its insulating effect and the lake floor freezes. As the permafrost advances, water trapped in sediments beneath the former lake is put under pressure. This pushes the overlying sediments into a dome-shaped hill or pingo.
- Open-system pingos originate when water trapped in a talik or in the active layer migrates under pressure through the frozen regolith. At a point of weakness, the water forces its way to the surface, forming a pingo.

Landscapes affected by thawing ground ice are known as **thermokarst**. Thermokarst contains thousands of shallow depressions that fill with meltwater to form **thaw lakes**.

Landforms of mass movement

Solifluction or **gelifluction sheets** are the most widespread mass movement landform in periglacial areas. They form vast expanses of smooth terrain, often at uniform angles as low as 1–3°. Large boulders known as **ploughing blocks** may be transported by solifluction/gelifluction sheets. They are 'rafted' downslope, their undersides resting at or near the permafrost table.

Lobes and **terraces** are common solifluction features. They give rise to step-like slopes with steep risers of 2 or 3 m and low-angled treads. For lobes to develop, solifluction/gelifluction must be concentrated into well-defined linear paths. Where movement is more uniform, terraces develop.

Some lobes and terraces have concentrations of large stones and boulders at their downslope ends. These are sometimes called **stone garlands** or **stone steps**. The stones appear to emerge from the terrace or lobe. This suggests that the sub-surface flow rates exceed those at the surface. **Turf-banked terraces** indicate that surface movement is greater. Angular boulders concentrated in valley bottoms by solifluction/gelifluction are known as **block streams**.

Asymmetric valleys

Asymmetric valleys in cross-section have one slope steeper than the other. They are common in periglacial regions. In the northern hemisphere, north-facing slopes tend to be steepest. The effect of aspect on the microclimate of slopes may explain asymmetry. Sun-facing slopes experience longer periods of thawing, more meltwater and greater solifluction/gelifluction. The result is a lowering of sun-facing valley slopes.

10 *Ecosystems in cold environments*

Cold environments support two types of tundra ecosystem: arctic and alpine. **Arctic tundra** is found polewards of the boreal (coniferous) forest in Eurasia and North America. It is treeless, poorly drained and dominated by low relief. Much of this region is underlain by permafrost. **Alpine tundra** occupies high mountain environments such as the Himalayas, Rockies and Andes. Both ecosystems support similar plant and animal species which are adapted to low temperatures, prolonged snow cover, strong winds, short growing seasons and poor soils (Table 4.1).

Table 4.1 Plant adaptations to tundra environments

Environmental problem	Adaptations
Short thermal growing season	Most plants are perennial. Perennials store food in tubers and rhizomes during the summer, which allows rapid growth to start in the spring.
	Some plants form flower buds a year in advance so as not to waste time during the short growing season.
	Many plants are evergreen so have no need to grow new leaves before photosynthesis can take place.
Low temperatures	Many plants have dark leaves to absorb the maximum insolation and raise their temperature.
	Many plants create their own microclimates by forming cushions, tussocks and rosettes.
	The parabolic shape of some flowers helps to concentrate on the developing reproductive parts. Other flowers track the sun on its path across the sky.
Strong winds	The low growing habit of most plants minimises resistance to the wind.
	Growth in cushions and tussocks reduces wind speeds by 99%.
	Many plants have waxy or leathery leaves that reduce moisture loss through transpiration.
Snow cover	Some plants complete their life cycles in a few weeks after they emerge from the snow. Plants that rely on food sources stored in tubers and rhizomes can survive under the snow for several years.

10.1 Environmental problems for plants in the tundra

Specific problems for plant adaptation in the tundra are:
- short thermal growing season
- limits to growth set by the hydrological growing season
- strong winds
- poor drainage and poor soils
- slow nutrient recycling
- frost heave and solifluction

Plant growth begins only when average daily air temperatures reach 6°C. As a result, much of the arctic and alpine tundra have a growing season lasting for just 30 to 60 days. Therefore, plants must flower, fruit and disperse their fruits in a very short time. Net primary production is low; averaged over the year it is around $0.01-0.66\,\mathrm{g\,m^{-2}\,day^{-1}}$.

Water is frozen for most of the year in the tundra. Therefore, even if temperatures are high enough for plant growth, water may be unavailable. Adding to these problems are strong winds and a lack of shelter. Strong winds cause physical damage to plants and increase moisture loss through transpiration.

During the summer when the surface layers of the soil thaw, the sub-soil remains frozen which creates waterlogged conditions. Elsewhere, free-draining soils are often **leached** and acidic and deficient in organic material and nitrogen. Nutrient cycling is inhibited by low temperatures and waterlogging, so that dead organic material is slow to decompose and **mineralise**. Meanwhile, frost heave and solifluction create unstable soil conditions, which further disrupts plant growth.

10.2 Food webs

Because of their low primary production, arctic and alpine ecosystems support animal populations of low density. Apart from mosquitoes, there are few insects and the climate is too cold for reptiles and amphibians. **Food chains** are short and **biodiversity** is low (Figure 4.12). As a consequence, tundra ecosystems are fragile and easily damaged by human activities and environmental change.

Some animals migrate to the arctic tundra in summer to breed. Barren ground caribou calve in the tundra in northern Canada to take advantage of the seasonal glut of food before migrating south to the coniferous forest for the winter. They are followed by predators such as the timber wolf. Many birds that spend the winter in mid-latitudes breed in the arctic tundra. A few mammals such as musk oxen and arctic foxes, which are protected by thick coats, remain in the tundra thoughout the winter.

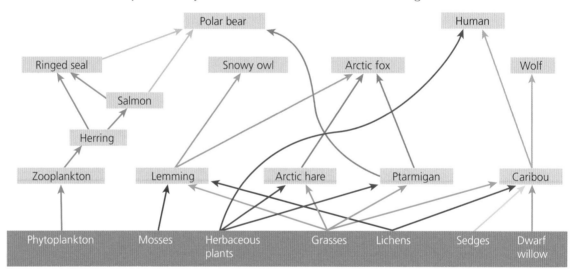

Figure 4.12 Arctic tundra food web

A feature of arctic-alpine animal populations is their 'boom and bust' character. Cyclic changes in food availability lead to population explosions of small rodents such as lemmings and pikas during years of glut. This results in a corresponding increase in predators like snowy owls and arctic foxes. Years of scarcity, however, lead to dramatic population crashes of both predator and prey species.

11 *Human activities in cold environments*

Cold environments provide opportunities as well as challenges for development. The economic opportunities include mineral and energy resource development (e.g. gas and oil, see case study) and exploitation of wilderness for recreation, leisure and **ecotourism**. However, development faces considerable challenges due to the severe climate, remoteness and ecological sensitivity of the arctic and alpine tundra. Development is further compromised by clashes with indigenous peoples such as the Inuit in northern Canada. Careful management is needed to ensure that tundra environments are not exploited for short-term gain and that development is sustainable.

CASE STUDY 1	The Arctic National Wildlife Refuge, Alaska
Issue	The Arctic National Wildlife Refuge (ANWR) is an important and fragile wilderness with protected status (Figure 4.13) on Alaska's north slope. However, the US government wants to develop oil and gas reserves in the ANWR. The potential impact of development on the environment, ecosystems and indigenous people could be disastrous. The proposed development is in conflict with the conservation status of the ANWR.

Figure 4.13 Arctic National Wildlife Refuge |
Background	The ANWR is a wilderness area covering 80,000 km². The ecosystem is pristine and has remarkable biodiversity. There are 130,000 caribou and 45 mammal species including wolves, wolverines, musk oxen and polar bears. The ecosystem is highly productive during the brief arctic summer and attracts large numbers of migrant birds, fish and caribou. There are two native Indian villages in the refuge that practise traditional subsistence economies. The indigenous people depend for survival on hunting whales, seals, walrus and caribou. In 2005 the US Senate gave permission for exploratory drilling for oil and gas in Area 1002 in the ANWR. The decision was the culmination of a long battle between oil companies and conservationists.
Potential impact	The US government gave the go-ahead for two main reasons: • to achieve greater security of energy supplies • to cut foreign energy imports and boost the USA's balance of payments Both are in the national interest. However, conservationists argue that drilling will disrupt the calving and migration of the caribou, permanently harm the traditional way of life and culture of indigenous people and damage the fragile tundra environment (oil spills, destruction of vegetation, melting of the permafrost). In addition, roads, pipelines and airstrips will have to be built on the tundra. Gravel will be extracted from stream and river beds for construction, adversely affecting freshwater ecosystems.
Management	The US government and oil companies say that modern technologies make oil and gas extraction less environmentally intrusive. Oil and gas reserves can be identified using controlled explosions, limiting the need for drilling and building access roads. Roads will be insulated from the permafrost on 'ice pads' and new drilling techniques allow oil and gas to be extracted several kilometres away from drilling rigs.

TOPIC 5 Coastal environments

1 The coastal system

The coast is an open system. Inputs of energy from waves, winds and tides interact with geology, sediments, plants and human activities to produce distinctive coastal landforms. The coast is a dynamic place where change (rockfall, landslides etc.) often occurs rapidly. This suggests that some parts of the coastal system have yet to achieve a steady state. Given that today's coastline is only 6,000 years old (i.e. the rise in sea level which followed the last glacial ended 6,000 years ago), this is not surprising.

2 Energy inputs

2.1 Waves

Wind waves are the main source of energy that drives the coastal system. Waves are superficial undulations of the water surface caused by winds blowing across the sea surface — a means of transmitting energy through water. Waves consist of orbital movements of water molecules that diminish with depth. In fact, water in a wave moves forward only when it approaches the shore and breaks. Wave power is proportional to the square of a wave's height and its velocity.

BOX 1 Wave characteristics

Wave length: the average distance between successive wave crests.

Wave height: the vertical distance between a wave trough and a wave crest.

Wave steepness: the ratio of wave height to wave length. Powerful waves are steep because they are high and have short wave lengths.

The **energy** in a wave is equal to the square of its height. Thus a wave that is 2 m high contains four times as much energy as a 1 m high wave. **Wave power** takes account of velocity as well as wave height. Thus:

Wave power = $H^2 \times V$

where: H = wave height
           V = wave velocity

Wave power is influenced by the openness of a coastline, the depth of water in the nearshore zone (Figure 5.1), the wind's strength and duration, and the length of open ocean over which the wave has been generated (**fetch**).

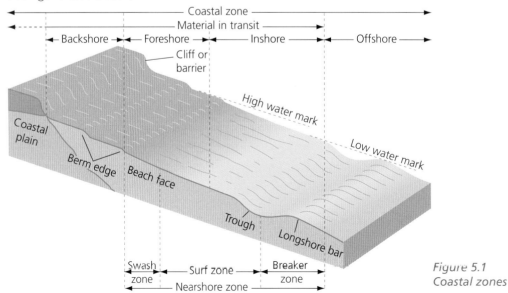

Figure 5.1
Coastal zones

- In high latitudes, open coastlines are usually high-energy environments dominated by erosional landforms.
- Coastlines sheltered from powerful waves (e.g. estuaries) are low-energy environments where tidal processes and deposition dominate.

2.2 Spatial and temporal variations in wave energy

The input of wave energy to a coastline varies in both space and time. Several factors influence wave energy:

- Fetch: the expanse of open water (in any direction) facing a coastline. Basically, the longer the fetch, the more powerful the waves. Along the coast of eastern England, northeasterly waves dominate coastal erosion and transport. In this direction there is the maximum fetch of around 2,000 km.
- Water depth: along shallow water coastlines (e.g. the Merseyside coastline), waves break some distance offshore and dissipate much of their energy.
- Wind strength: the stronger the wind, the more energy transferred to the waves. Gale force winds generate damaging storm waves.
- Wind duration: a steady breeze, blowing for several hours or days, transfers huge amounts of energy to the sea surface, and can generate large waves.

Landforms of coastal deposition

Beaches are accumulations of sand and shingle deposited by waves in the inshore zone (Figure 5.2). The sediments that form beaches come from three sources: cliff erosion, offshore areas and rivers.

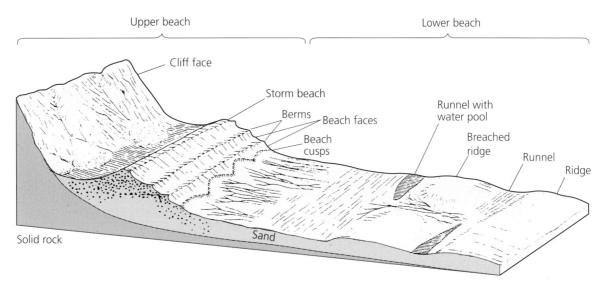

Figure 5.2 Principal features of a sand and shingle beach

- Cliff erosion: on average, cliff erosion provides only a small fraction of beach sediments (around 5%).
- Offshore areas: some beach sediments have been combed from the shallow seabed. This process occurred since the end of the last glacial. Between 18,000 years and 6,000 years ago, sea level rose by around 100 m. Coarse alluvium deposited by rivers on the exposed continental shelf was swept shorewards by wave action, providing sediment for today's beaches.
- Rivers: rivers are the source of around 90% of beach sediments. Sand and shingle are transported into the coastal system through river mouths as bedload.

Coastal sediments are confined to well-defined stretches of coastline known as **sediment cells** or **littoral cells**. The major sediment cells in England and Wales are the basis of coastal management (Figure 5.3).

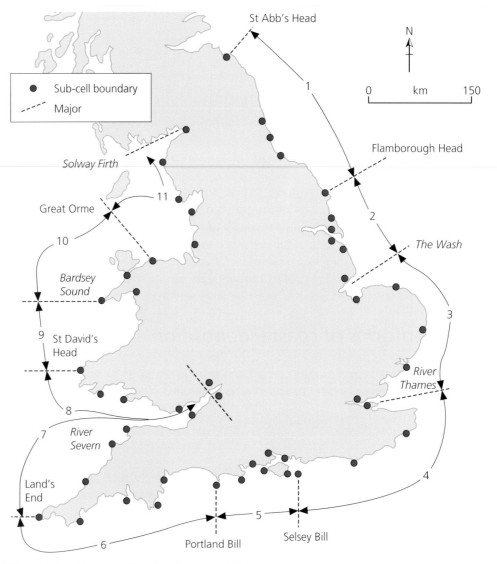

Figure 5.3 Coastal sediment cells in England and Wales

3.1 Beach profiles

The cross-section of a beach between the mean high water mark and the mean low water mark is known as the **beach profile**. Two factors influence beach profiles: sediment size and wave type.

Sediment size

Shingle beaches are usually steeper and narrower than sand beaches. Shingle is coarser than sand and has a higher **percolation rate**. The swash pushes shingle up the beach, but with rapid percolation it soon loses power and there is little or no backwash to drag the shingle seawards. Therefore, the shingle is moved only in one direction. The outcome is a beach with a relatively steep angle (e.g. about 10°). Sand beaches, with lower percolation rates, have a longer swash and more powerful backwash. The result is a beach with a lower average slope angle.

Wave type

Beaches made of similar sediments often have different profiles. This demonstrates the influence of wave type. High-energy surfing breakers produce wide, flat beaches. These waves have a powerful backwash and erode sand and shingle from beaches. The sediments are transferred offshore, where they form a **longshore bar**. The resulting wide, flat beaches and longshore bar absorb wave energy until there is no net onshore/offshore sediment transport. Low-energy waves or surging breakers induce a net onshore transfer of sediments. As a result, beaches become steep, with prominent **beach faces** and **berms** (Figure 5.2).

3.2 Beach plans

In planform, beaches are either swash-aligned or drift-aligned.

Swash-aligned beaches

Crescent-shaped bay-head beaches develop on indented coasts where waves are fully refracted (Box 2). **Swash-aligned beaches** are usually straight, without the recurved laterals that indicate **longshore drift**.

Box 2 *Wave refraction*

Wave refraction describes the bending of oblique waves in the nearshore zone until they break parallel to the shore. Where the sea is shallow, waves 'feel' the seabed and slow through frictional drag. In deeper water, the wave moves faster. As a result, the wave front bends until it takes on a similar shape to the coastline. When the waves are fully refracted and parallel to the coastline, the swash and backwash follow the same path on the beach. The result is swash-aligned beaches.

Even so, not all waves are fully refracted. Some waves break obliquely

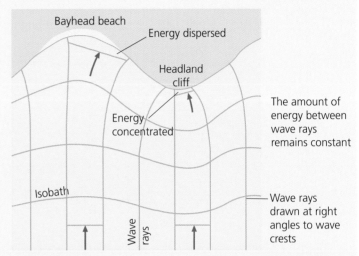

Figure 5.4 Wave refraction and the distribution of energy along coastlines

and their swash follows a diagonal path across the beach. In these circumstances, a net lateral movement of beach sediment takes place. This is longshore drift and it produces drift-aligned beaches such as spits.

Wave refraction is also responsible for the uneven distribution of energy on coastlines. Refraction concentrates wave energy (and therefore erosion) on headlands, but disperses energy (favouring deposition) in bays. The process of wave refraction is shown in Figure 5.4.

Box 3 *Chesil Beach*

At 30 km long, Chesil Beach (Figure 5.5) is the longest shingle ridge in the British Isles. It joins mainland Dorset with the Isle of Portland. Such a beach, linking the mainland to an island, is known as a **tombolo**. Because of its straightness, Chesil is thought to be a swash-aligned feature. It originated during the last glacial as a shingle bar of flint and chert in the English Channel. At that time sea level was 100–120 m lower than today. Rising sea levels followed the end of the glacial and waves gradually rolled the bar onshore. Chesil reached its present position 6,000 years ago.

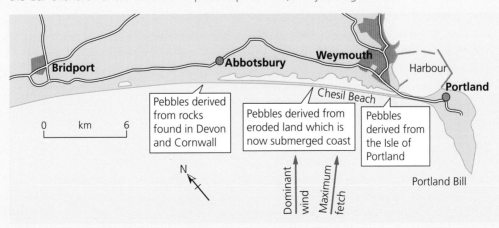

Figure 5.5 Chesil Beach, west Dorset

Drift-aligned beaches

Drift-aligned beaches such as **spits** and **barrier beach islands** develop on open coastlines. Here, waves are rarely fully refracted and sediment transport is dominated by longshore drift (Figure 5.6). Spits often form across estuaries (e.g. Spurn Head, east Yorkshire; Orford Ness, Suffolk) or where there is an abrupt change of direction in the coastline (e.g. Hurst Castle, Hampshire). Growth by longshore drift is shown by the **recurved** shingle ridges or laterals of spits. The distribution of spits around the UK appears to correlate with coastlines that have a low tidal range (less than 2 m). A low tidal range concentrates wave action in a narrow vertical band of coast. This seems to be important in shaping the sand and shingle into spits and other drift-aligned beaches.

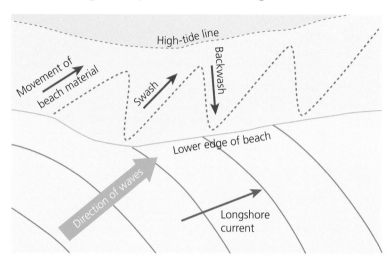

Figure 5.6 Longshore drift: the flow of coastal sediment produced by wave and current action when waves approach at an angle to the coastline

Box 4 *Blakeney Point and Scolt Head Island*

Blakeney Point is a spit 15.5 km long on the north Norfolk coast (Figure 5.7). Unusually, it is situated on a relatively straight coast and aligned at a steep angle to the coast. It has a series of recurves that form prominent shingle ridges. These mark the westward growth of the feature by longshore drift. However, there is little evidence of longshore drift today. The steep angle that the spit makes with the coast and the gradual movement of the feature inland suggest that Blakeney Point may now be swash-aligned.

Scolt Head Island, also on the north Norfolk coast, is similar to Blakeney Point in plan, except that at low tide it is separated from the mainland by a deep narrow channel. This makes Scolt Head a barrier beach island. Like Blakeney, Scolt Head's recurves show that it has formed by longshore drift. As at Blakeney, Scolt Head seems to be gradually moving onshore. The source of shingle at Blakeney and Scolt Head Island is glacial gravel. Deposited offshore by meltwater, it was swept by waves towards the modern coastline as sea levels rose in post-glacial times.

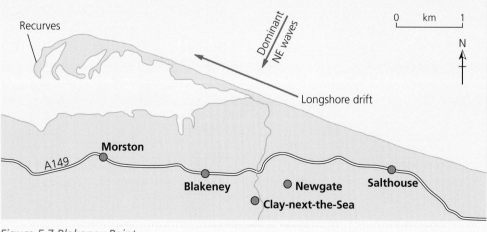

Figure 5.7 Blakeney Point

4 | *Mudflats and salt marshes*

Mudflats and **salt marshes** are landforms of sheltered, low energy coastlines where wave action is weak. In quiet areas such as estuaries, and on the landward side of spits, tidal currents deposit fine sediment in suspension. Accretion leads to the growth of mudflats and salt marshes.

Unlike beaches, mudflats and salt marshes are often associated with large tidal ranges. Large tidal ranges generate powerful tidal currents that can transport large quantities of fine sediment.

Figure 5.8 is a cross-section through a salt marsh in the Aln estuary in Northumberland. It shows a number of typical features:
- There is a low cliff, about half a metre high, which separates the mudflats from the low marsh. Tidal scour maintains this cliff.
- An abrupt break of slope separates the high marsh from the low marsh.
- A dense network of creeks and tributaries drains the marsh and brings water in on the flood tide.
- Small salt pans occur sporadically on the surface of the high marsh.

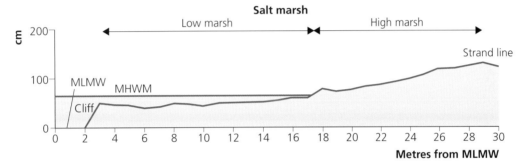

Figure 5.8 Cross-section through a salt marsh at Alnmouth, Northumberland

Vegetation plays a vital role in the development of salt marshes. Many marshes demonstrate a zonation of species that is closely related to height above sea level, as shown in Table 5.1. Such a plant succession in a saline, waterlogged environment is known as a **halosere**. It is best seen between the mean low water mark and the strand line on open marshes in estuaries.

Table 5.1 Plant zonation on salt marshes

Environment	Environmental conditions	Plants
Mudflats	High salinity levels; low oxygen levels in mud; high turbidity; long periods of inundation on each tidal cycle.	No plants, only algae.
Low marsh	Less hostile conditions than mudflats, but salinity and turbidity still high and oxygen levels low. Tidal inundation shorter.	*Spartina* (cord grass) and *Salicornia* (glasswort) are the two common pioneer species. Sea blite (*Suaeda maritima*) and sea purslane (*Halimone portulacoides*) on better-drained areas (e.g. edges of creeks).
High marsh	Flooding only occurs on spring tides. Salinity levels are relatively low and soil develops.	Wide variety of species, including salt marsh grass (*Puccinellia*), sea rush (*Juncus maritimus*), sea lavender (*Limonium*), sea aster (*Aster tripolium*), sea blite (*Suaeda maritima*) and sea purslane (*Halimione portulacoides*).

Ecological succession is responsible for the growth of salt marshes and follows a number of stages.
- Colonisation by pioneer species such as cord grass and glasswort. These plants slow the movement of water and encourage rapid sedimentation (1–2 cm a year). Their roots help to stabilise the mud.
- Through accretion of sediment the marsh increases in height and conditions become more favourable for the invasion of other, less tolerant species. Biodiversity and plant cover increase. Plants such as sea rush, sea aster, sea lavender, salt marsh grass and common scurvy grass dominate.

● The marsh height stabilises a metre or so above the mean high-tide mark. With only occasional inundation on the highest spring tides, vertical accretion ends. Salinity levels have dropped and soil starts to develop.

5 *Sand dunes*

Sand dunes are the only important coastal landforms produced by the wind. The wind induces the movement of sand by the collision of particles. Larger particles move by **creep** on the surface of dunes; smaller particles are transported by a skipping process that extends up to a metre or so above the surface. This is **saltation**.

Blown sand accumulates around objects such as logs and bottles. However, once these are buried, sand accumulation ends. Sand dunes can only form when vegetation provides the obstacle to blown sand. This is because some plants, such as marram grass (*Ammophila arenaria*), thrive when submerged by sand. Burial stimulates rapid growth, which in turn encourages further deposition. Deposition also occurs because plants reduce wind speeds near the ground surface.

Sand dunes develop on lowland coastlines where the following conditions are found:
● a plentiful supply of sand (possibly from nearby river estuaries)
● a shallow offshore zone with gentle gradients, where large exposures of sand dry out at low tide
● an extensive backshore area where sand can accumulate
● prevailing onshore winds

The primary succession on sand dunes (**psammosere**) often shows a clear zonation. Nearest the shore, where the youngest (yellow) dunes are found, vegetation cover is patchy and few species survive. With increasing distance inland, the dunes get older (grey dunes) and the environment changes (Table 5.2). As a result, plant cover, productivity and biodiversity all increase.

Table 5.2 Vegetation succession on sand dunes

Dunes	Environmental conditions	Plants
Embryo dunes	Little fresh water for plants. Sand is extremely porous. Blown sand blasts plants. Salt spray and shifting sand are further problems for plants.	Sand twitch or sea couch grass (*Agropyron junceiforme*), which can extract fresh water from salt water, is one of the few species found here.
Foredune ridge	Two or three metres above the beach, conditions are less saline, more exposed and unstable.	Marram grass begins to colonise and compete with sea couch grass. Marram is well adapted to sand dunes: it has deep roots, sand deposition stimulates growth and moisture loss is reduced by a thick shiny cuticle, sunken pores and leaves that curl in the sun.
First dune ridge	The most prominent relief feature, 10–30 m high. Sand is more stable and contains some humus.	Almost 100% marram. Marram slows wind speed and reduces sand movement.
Older dune ridges	Dune ridges become progressively lower inland as sand supply diminishes. Dunes become grey as humus builds up in the soil. Soil's moisture retention increases. Soil pH falls.	Environment becomes less harsh. There is shelter, some fresh water and soil. Many new species appear, e.g. creeping willow, sea buckthorn, ragwort, fescue grass. Marram dies out.
Dune slacks	Depressions between the dune ridges. Sheltered and marshy where water table reaches the surface. Often waterlogged after heavy rain.	Wide range of plant species. Some, like flag iris and bog myrtle, are typical wetland species.

Coastal dunes form a series of ridges parallel to the coastline, as shown in Figure 5.9. The dunes decrease in height inland as sand supply diminishes. Depressions or **slacks** separate the dune ridges. There the water table reaches the surface and, in contrast to the arid dunes, creates a wetland habitat. Coastal dunes usually have a steep windward slope and a less steep leeward slope. Sand eroded from

the exposed windward slopes is deposited on the sheltered leeward side. The older dunes are immobile, stabilised by their vegetation cover. Where supplies of fresh sand are plentiful, dune systems extend seawards. This process, known as **progradation**, produces parallel dune ridges.

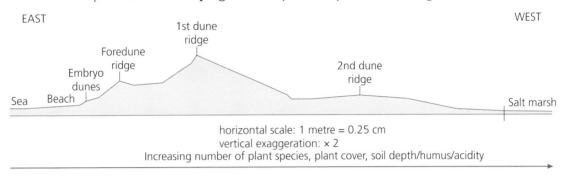

Figure 5.9 Cross-section through the dunes at Alnmouth

Sand dunes are fragile environments. Destruction of the vegetation cover by grazing (e.g. rabbits) or human activities (e.g. trampling, firing) can lead to massive wind erosion and the formation of **blow-outs**. Frontal erosion of dunes by waves also occurs in storm conditions. Rising sea levels due to global warming will be an increasing threat to coastal dune environments in the twenty-first century.

6   *Landforms of coastal erosion*

6.1 Marine erosional processes

Wave action on coasts induces three erosional processes: abrasion (or corrasion), hydraulic action (or quarrying) and corrosion (Table 5.3). These processes are most important when high-energy waves, associated with storm conditions, strike coasts made of weak and incoherent rock.

Table 5.3 Marine erosion processes

Type of erosion	Process
Abrasion (corrasion)	High-energy waves pick up shingle and scour (abrade) the base of cliffs. The result is a wave-cut notch. The cliff is undermined and retreats through rockfall.
Hydraulic action (quarrying)	Air and water, forced under pressure into joints and bedding planes by storm waves, weaken rocks and cause collapse. The effectiveness of hydraulic action depends on the density of joints etc. in the rock. Hydraulic action also includes the impact of masses of water (wave shock or wave hammer), which loosens and dislodges rock particles.
Corrosion	Some rock minerals are susceptible to solution. For example, calcareous cements that bind sandstone particles may be attacked by solution, leading to rock disintegration.

6.2 Cliff profiles

Cliff profiles (i.e. cross-sections) reflect the influence of rock lithology and structure, sub-aerial processes, relief, wave energy, human activities and past processes.

Lithology and structure

Mechanically strong rocks, such as basalt and limestone, not only resist erosion but are stable at steep angles. In contrast, mechanically weak rocks, such as clay and shale, will erode more easily and often form lower-angled slopes.

Rock structure includes the angle of dip of sedimentary rocks. The effect of lithology on cliff profiles is shown in Figure 5.10.

- Horizontally bedded rocks, such as the chalk at Flamborough Head in Yorkshire and at Bat's Head in Dorset, form vertical cliffs. These cliffs, undermined by wave action, retreat parallel to themselves, maintaining their steep slope by rockfall.

- Landward-dipping strata also form steep cliffs. This is because eroded and weathered rock particles are not easily dislodged from the cliff face.
- Seaward-dipping strata develop profiles that mirror the angle of dip of the bedding planes. Blocks weakened by erosion and weathering fail along these planes and slide into the sea.

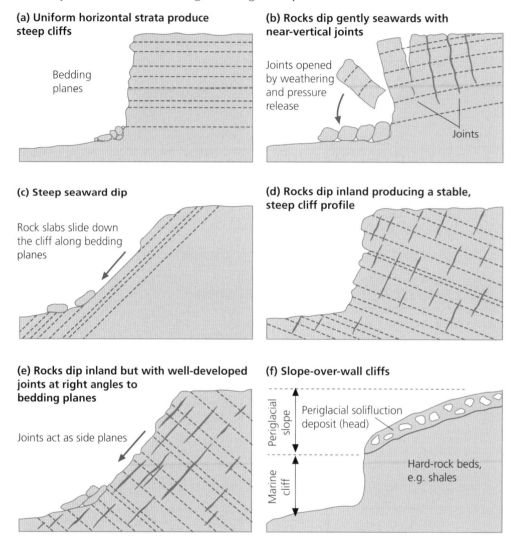

(a) Uniform horizontal strata produce steep cliffs

Bedding planes

(b) Rocks dip gently seawards with near-vertical joints

Joints opened by weathering and pressure release

Joints

(c) Steep seaward dip

Rock slabs slide down the cliff along bedding planes

(d) Rocks dip inland producing a stable, steep cliff profile

(e) Rocks dip inland but with well-developed joints at right angles to bedding planes

Joints act as side planes

(f) Slope-over-wall cliffs

Periglacial slope

Marine cliff

Periglacial solifluction deposit (head)

Hard-rock beds, e.g. shales

Figure 5.10 Cliff profiles: the influence of lithology

Some cliffs comprise two or more rock types of contrasting lithology and structure. On England's northeast coast, boulder clay caps many hard rock cliffs. Such cliffs have two distinctive slope elements: a steep lower cliff wall cut into solid rock and a more gentle upper slope of boulder clay.

Sub-aerial processes

Sub-aerial processes such as weathering and mass movement also have a strong influence on cliff profiles. This is particularly evident in weak rocks such as shale and sands, which are susceptible to slides, flows, surface wash and gulleying. Weathering processes in coastal environments include wetting and drying due to waves, spray and tides; and salt weathering. Honeycomb rock surfaces in the spray zone are evidence of salt weathering. Rocks on the shoreline are also destroyed by biological weathering caused by organisms such as molluscs, sponges and sea urchins.

Relief

Relief determines the height of cliffs and the area over which sub-aerial processes operate. In County Clare, the coast rises above 300 m to form the Cliffs of Moher, the highest in the British Isles. In contrast, in Christchurch Bay on England's south coast, low relief and tertiary sands and clays support cliffs with a modest height of between 10 and 30 m.

Wave energy

- Coastlines with long fetches or exposed to prevailing winds (e.g. east coast of Hawaii exposed to the northeast trade winds) experience high wave energy. As a result, they tend to have relatively high rates of erosion. Moreover, rockfalls such as the one at Beachy Head in 1999 are easily removed by strong wave action so that erosion can begin anew.
- Low-energy coastlines often experience relatively low rates of erosion. Waves may have low energy owing to the coast's sheltered situation, short fetch and shallow water offshore (the last dissipates wave energy before it reaches the coastline). In low-energy environments, sub-aerial processes begin to dominate. Rock debris may accumulate at the cliff foot and cliff slope angles may be lowered by weathering and mass movement.

Human activities

Human activities may either increase or decrease rates of marine erosion.

- As part of coastal management, cliffs may be protected from erosion by coastal defence works (sea walls, revetments etc.). In this situation, marine erosion stops, cliffs become vegetated and sub-aerial processes take over. At Mappleton in Holderness, the boulder cliff angles have been deliberately lowered to reduce the risk of slope failure.
- Human interference in the coastal system has sometimes (inadvertently) accelerated erosion. The mining of sand and shingle from offshore banks can disrupt the local sediment budget. The result is a thinning and narrowing of beaches, leading to accelerated cliff erosion. This happened in the early twentieth century at Hallsands in south Devon and resulted in the destruction of the village.

Past processes

Many cliff profiles owe their shape to past processes. Slope-over-wall or bevelled cliffs have a long, convex upper slope and a short, vertical lower section. They are common on many coastlines, especially in parts of south Devon and south Cornwall. The long, convex slope is unrelated to marine erosion. It formed under cold climatic conditions during the last glacial. At this time, processes such as freeze–thaw and solifluction prevailed, and sea level was 100–120 m lower than today. Only the basal part of this convex slope has been trimmed back (to form a 'wall') by wave action in the last 6,000 years.

6.3 Erosional landforms resulting from cliff recession

Landforms of cliff erosion and recession on upland, hard rock coasts include caves, arches, stacks, blow holes, geos and shore platforms.

- **Caves** develop below the mean high water mark along lines of weakness such as joints, bedding planes and faults. Hydraulic action and abrasion loosen blocks along joints to create hollows that can be further exploited by waves.
- Caves that form on opposite sides of headlands (where wave energy, owing to refraction, is highest) may form **arches** such as Durdle Door and the Green Bridge of Wales.
- The combination of marine and sub-aerial processes leads to arch collapse (as at Marsden in Tyne and Wear in 1996), leaving isolated rock pinnacles or **stacks** (e.g. the Needles on the Isle of Wight).
- If part of the roof of a tunnel-like cave collapses along a master joint, it may form a vertical shaft that reaches the cliff top. This is a **blow hole**.
- When the entire roof of a cave running at right angles to the cliff line collapses, it forms a narrow inlet or **geo**. A natural arch may also develop if part the cave roof remains intact at the seaward end of a geo (e.g. at Flamborough).
- The final stage of cliff recession results in the ultimate erosional feature: a wide shore platform. This, the former base of the cliffs, is abraded by wave action and weathered by biological and chemical processes. The profile of shore platforms is linked to the tidal range (Figure 5.11). Coastlines with a low tidal range (less than 4 m) have gently sloping platforms interrupted by a ramp at the high-tide mark and a small cliff at the low-tide mark. Erosion is concentrated at these two locations during the tidal cycle. On coastlines with a large tidal range, erosion is spread over a wider area of the platform. This creates a more uniform and steeper sloping platform.

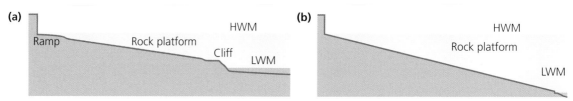

Figure 5.11 Shore platform

7 Rock structure and the planform of coasts

At a regional scale, rock structure has a strong influence on the shape of coastlines.
- Where the main rock types crop out parallel to the coast, the coast is often straight and uniform. We call such coastlines **accordant** or Pacific coasts.
- Where rocks of different types crop out at right angles to the coast, the result is a coastline of headlands and bays. The more resistant rocks form headlands; the less resistant ones form bays. This form of coastline is known as the **discordant** or Atlantic type.

The coastline of Purbeck in southeast Dorset, shown in Figure 5.12, illustrates both accordant and discordant types. In south Purbeck, the principal rock types — Portland limestone, Purbeck limestone, Wealden beds and chalk — run parallel to the coast. The Portland limestone is the southernmost outcrop. It forms a barrier to erosion and is responsible for the straightness of the coastline. However, in some places such as Stair Hole, erosion by waves and rivers has breached the Portland barrier and exposed the weaker Wealden beds. These weaker rocks have been carved into impressive coves and bays. The best examples are Lulworth Cove and Worbarrow Bay.

The east Purbeck coast has the same geology as south Purbeck. The difference is that in east Purbeck the rocks crop out at right angles to the coastline. Therefore, the resistant Portland and Purbeck limestones and chalk form headlands such as Durlston Head and Peveril Point, while the weaker Wealden beds and Bagshot sands erode to form wide bays like Swanage and Studland.

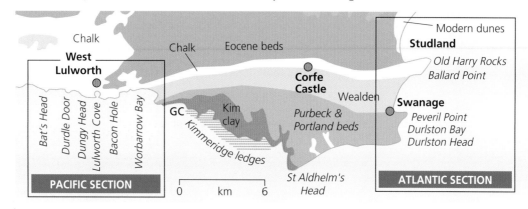

Figure 5.12 The Isle of Purbeck, southeast Dorset

8 Sea level change

Absolute sea level changes result from a worldwide rise or fall in the volume of water in the oceans. This is a **eustatic** change. During the last glacial, sea level was 100–120 m lower than today and large areas of the continental shelf around the British Isles were dry land.

Rising sea level, which causes the coastline to retreat, is called a **transgression**. This happens during interglacial periods, when ice sheets and glaciers melt. The opposite, **regression**, produces an advancing coastline. This occurs during glacials when huge volumes of water are locked up in ice sheets and glaciers. For example, in the last glacial, ice sheets and glaciers in the northern hemisphere occupied an area three times greater than today.

At a local scale, sea level change occurs if the land rises or sinks relative to sea level. This type of sea level change is associated with either **isostatic** or **tectonic** movements. Table 5.4 summarises sea level changes and their effects on coastal landforms.

Table 5.4 Changing sea levels and coastal landforms in the British Isles

Glacio-eustacy	
Rising sea level (absolute — submergence)	
Shingle beaches (spits, bars etc.)	River sediments deposited on the dry continental shelf during the last glacial when the shelf area was above sea level. Rising sea level in the last 20,000 years swept up the sediment and deposited it on present-day coasts.
Estuaries	Drowned, shallow lowland river valleys, e.g. Severn, Thames and Humber estuaries. Transgression flooded low-lying areas around the Wash and Somerset Levels.
Rias	Drowned, incised river valleys on upland coasts, e.g. southwest Ireland, River Dart (Devon), River Fal (Cornwall).
Fjords	Drowned glacial troughs.
Ancient shore platforms	Cut by wave action in the last interglacial when sea level was 8–10 m higher than today, e.g. at Start Bay in south Devon.
Glacio-isostacy	
Falling sea levels (relative — emergence)	
Raised beaches and relict cliffs	Ancient beaches and cliff lines elevated above sea level following deglaciation and the unloading of ice sheets from northern Britain, e.g. Applecross Peninsula, northwest Scotland (8 m above sea level).
Tectonic movements	
Earthquakes/ faulting	Localised tectonic movements leading to submergence or emergence, e.g. Alaskan earthquake in 1964 raised part of the Alaskan coast by 4–9 m.

Box 5   *Isostatic change*

During a glacial, the great mass of ice sheets and glaciers loads the continental crust, causing it to sink by several hundred metres. When the ice melts, unloading occurs and the crust slowly rises by a similar amount. This process is called **glacio-isostacy**. In the British Isles, the ice was thickest in Scotland and it is here that isostatic change has been greatest. This movement, which is ongoing, has taken place faster than the post-glacial rise of the sea level, creating raised beaches and raised shorelines around Scotland's west coast.

9   *Coastal management*

Most of the world's population lives on or near the coast. The coast provides a wide variety of opportunities for recreation and leisure, industrial location, trade, wildlife, conservation, land reclamation, farming and waste disposal.

Because of the importance of coasts to human activity, management of the coastal environment to control erosion, flooding, overdevelopment and pollution is common.

9.1 Shoreline management plans

In England and Wales, shoreline management to address problems of erosion and flooding are the responsibility of local authorities (erosion) and the Environment Agency (flooding). Both problems are coordinated through shoreline management plans (SMPs).

Modern coastal management views the coast as an integrated system. This means that coastal resources, their exploitation and development, and issues such as erosion and flooding, should be managed together rather than piecemeal. Such an approach recognises the interrelatedness of natural and human systems and that change to one part (e.g. stopping erosion) is likely to have adverse knock-on effects elsewhere.

The basic unit of coastal management in England and Wales are the eleven sediment cells (see Figure 5.3 on page 48). Within these 'natural units' the inputs and movements of silt, sand and shingle are self-contained. SMPs set out a strategy for coastal defence for specified lengths of coast identified within sediment cells (i.e. sub-cells or management areas).

Two important ideas dominate modern thinking on coastal management:
● human intervention in the natural processes that operate in the coastal zone should be minimal
● where intervention is necessary, it should be sustainable

In the past, human intervention in the coastal sediment system has often been disastrous (Figure 5.13 and Table 5.5). Meanwhile, the need for sustainable intervention is made more urgent by global warming. Rising sea levels and increased storms will place even greater pressure on coastal defences in future.

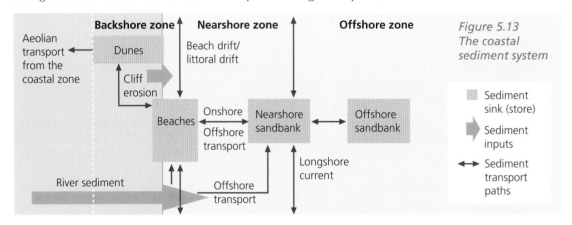

Figure 5.13
The coastal sediment system

Table 5.5 Coastal protection: hard and soft engineering responses

Hard engineering

Sea walls	Expensive to build and maintain. Building justified only to defend settlements of some size. Sea walls reflect wave energy. This sets up vertical currents that can undermine and topple sea walls. Sea walls protect the shoreline but may cause accelerated erosion of beaches. They also reduce sediment input to the coastal system by natural erosion and contribute to the narrowing and thinning of beaches.
Revetments	Wooden or rock barriers parallel to the coast designed to absorb wave energy. Cheaper than sea walls, but unsightly. They also limit access to beaches by visitors and tourists.
Groynes	Wooden or rock barriers at right angles to the shore. They interrupt the longshore movement of sand and shingle. By keeping the beach intact, they reduce erosion. However, they may starve beaches downdrift of sand and shingle and accelerate erosion there.
Rock armour	Boulders or concrete blocks placed at the foot of cliffs or in the backshore area of a beach. Unsightly, but cheap and effective.

Soft engineering

Beach replenishment	Sand and shingle brought to beaches to replace sediments lost to longshore drift. Beaches are effective absorbers of wave energy, providing the sediment remains *in situ*. Problems may occur if sand for replacement is mined offshore, disrupting the coastal sediment system.
Managed realignment	Existing hard defences may be dismantled or left to fall into disrepair. The shoreline is set back, allowing the sea to flood areas previously protected. Salt marshes, mudflats and beaches develop on the newly flooded areas and provide natural protection against erosion and flooding. This approach is sustainable and has significant environmental benefits (e.g. new intertidal habitats for wildlife).

Coastal management in SMPs comprises three strategies:

● **hold the line:** aims to maintain and in some cases strengthen existing coastal defences. This policy is justified where the value of properties and infrastructure at risk from erosion and flooding exceed the cost of defences

● **no active intervention:** covers most of the coastline of England and Wales. Natural processes are allowed to operate without human interference. Sometimes conflicts arise if properties and parts of settlements may be lost to coastal erosion and flooding (e.g. at Happisburgh in northeast Norfolk).

● **managed realignment:** applies to only a few stretches of coastline. It involves setting back the shoreline and allowing the sea to flood areas that were previously protected by embankments and sea walls. The motives for managed realignment include:
  — reducing the costs of maintaining non-sustainable or expensive sea defences
  — allowing salt marshes and mudflats to develop on abandoned farmland to create natural defences against flooding and erosion
  — creating new wildlife habitats such as intertidal mudflats and salt marshes. For example, in Essex 75 % of coastal salt marshes have been drained and used for arable production in the last 100 years
  — tackling the problem of coastal squeeze (Box 6)
  — restoring sediment sinks

Box 6 *Coastal squeeze*

Coastal squeeze describes the erosion of beaches and salt marshes caused by a combination of hard coastal defences, rising sea levels and reduced supplies of sediment. It occurs when beaches and salt marshes are trapped between rising sea level and hard defences such as sea walls and flood embankments.

On natural coastlines, beaches and salt marshes would simply migrate inland. Hard defences make this impossible and as a result beaches and salt marshes are eroded.

Scheme	Description
Mining of sand and shingle	Mining of sand and shingle for aggregate for use in construction and/or beach replenishment reduces the total sediment supply. As sediments often move from beaches to offshore bars and vice versa, beaches are frequently starved of sediment. As a result, they become thinner and narrower and coastal erosion accelerates. For example, the mining of sand from the Dolphin/Shingles offshore bars in Christchurch Bay has accelerated the erosion of Hurst spit.
Building groynes, piers and jetties	These structures interrupt the longshore movement of sediment. Beaches downdrift of groyne fields are starved of sediment and the coastline in this area may suffer accelerated erosion (e.g. the loss of the farm at North Cowden following the construction of groynes in the early 1990s at Mappleton on the Holderness coast).
Sea walls and hard defences	Sea walls protect the coastline from erosion and therefore reduce sediment inputs into the coastal system. In the long term this will diminish sediment stores such as beaches and offshore bars. Sea walls and other hard defences are partly responsible for coastal squeeze and the erosion of beaches, salt marshes and mudflats.
Dams and weirs	The construction of dams and weirs on rivers trap sediments and reduce sediment supplies to the nearshore zone. The erosion of the Nile and Yangtze deltas is due to the building of the Aswan and Three Gorges dams respectively.

Table 5.6 Human impacts on the coastal sediment system

9.2 Shoreline management in Christchurch Bay

Christchurch Bay in Hampshire (Figure 5.14) lies between Hengistbury Head and Hurst Spit. It is a dynamic coastline with rapid erosion, sediment transport and deposition. Rapid erosion is due to cliffs made of weak sands and clay, powerful waves from the southwest and little fresh sediment input (largely caused by coastal defence structures). Current rates of erosion at Barton-on-Sea are nearly 2 m a year. The coastline is densely populated and includes the urban areas of Christchurch, Barton-on-Sea and Milford. As a result, coastal management is a sensitive issue.

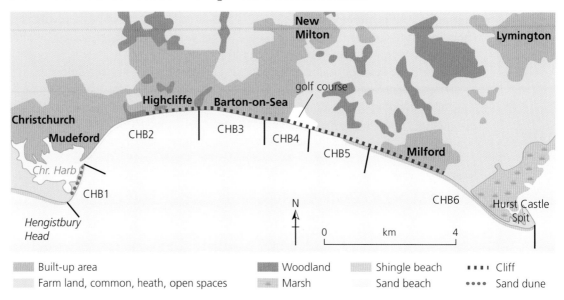

Figure 5.14 Christchurch Bay, Hampshire

The SMP includes all three management strategies for the coastline: hold the line, no active intervention and managed realignment. These policies are defined for the six management units that make up Christchurch Bay (Table 5.7).

Table 5.7 SMP policies for Christchurch Bay

Management unit	Policy
CHB1	Realign and hold existing line
CHB2	No active intervention
CHB3	No active intervention
CHB4	Hold the line
CHB5	No active intervention
CHB6	Hold the line

1 Distribution of hot arid and semi-arid environments

Nearly a third of the Earth's land surface is arid or semi-arid. Hot arid and semi-arid environments occupy the arid zones in the topics and sub-tropics. Aridity is defined by the ratio of precipitation (P) to **potential evapotranspiration** (PET). Potential evapotranspiration is the amount of moisture which, if available, would be removed from the land's surface by evaporation and transpiration. Everywhere in the hot arid and semi-arid zone annual potential evapotranspiration exceeds precipitation. Table 6.1 shows that potential evapotranspiration is more than 97 % higher than precipitation in hyper-arid environments, 20–97 % higher in arid environments and 20–50 % higher in semi-arid environments.

Most arid lands are found between latitudes of 20 and 35°N. In the northern hemisphere, the hyper-arid and arid zone is mainly accounted for by the Sahara, Arabian and Gobi deserts. In the southern hemisphere, the arid zone includes most of Australia, southwest Africa, southern Argentina and northern Chile (Figure 6.1).

Table 6.1 Classification of arid lands

	P/PET	Mean annual precipitation (mm)	% of world's total land area
Hyper-arid	< 0.03	< 100	4.2
Arid	0.03–0.2	100–300	14.6
Semi-arid	0.2–0.5	300–600	12.2

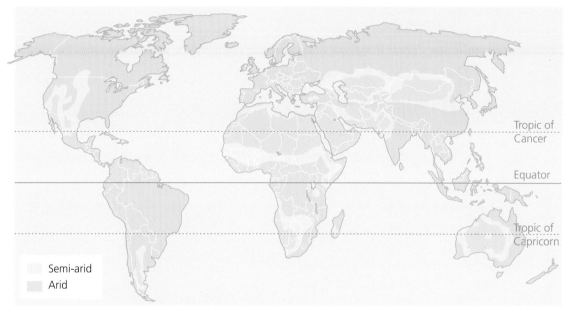

Tropic of Cancer

Equator

Tropic of Capricorn

Semi-arid
Arid

Figure 6.1 The distribution of arid environments

Box 1 *Aridity and rainfall effectiveness*

Rainfall effectiveness is defined as the amount of rain that reaches the root zone and is available to plants. It is calculated as follows: *Rainfall effectiveness = Actual precipitation − Actual evapotranspiration*

A combination of factors affect rainfall effectiveness:

■ **Rates of evaporation**: evaporation is a function of temperature and wind speed. In hot **drylands**, a large proportion of rainfall is lost to evaporation.
■ **Seasonality**: winter rainfall has greater effectiveness than summer rainfall because evapotranspiration losses are lower in winter.
■ **Rainfall intensity**: rain falling in heavy convectional storms runs off quickly and there is little **infiltration** into the soil.
■ **Soil type**: clay soils have limited capacity to absorb water and therefore create additional runoff. Sandy soils are porous and are highly susceptible to drought. Clay soils baked by the sun also reduce infiltration and accelerate runoff.

2   *The causes of aridity*

Aridity in the tropics and sub-tropics has several causes:
- the general circulation of the atmosphere between the equator and latitude 30–35°N
- topography
- cold ocean currents

2.1 General circulation

Two large convective cells known as **Hadley cells** control the atmospheric circulation and climate in the tropics (Figure 6.2). Around the equator, intense **insolation** creates powerful convection currents. Warm air rises 10–15 km above the surface and then moves polewards. As it drifts to higher latitudes it cools, and then, at around latitudes 20–30°N, it sinks towards the surface. Warmed by compression, this sinking results in cloud-free conditions and forms a large permanent **anticyclone** at the surface. This process is responsible for the formation of the world's largest deserts (e.g. Sahara, Arabian and Great Victorian).

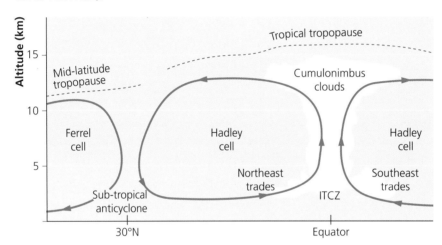

Figure 6.2 Hadley cells

2.2 Topography

Mountain ranges create **rain shadows**, which in extreme cases cause desert environments. Air masses crossing mountain ranges are cooled and shed much of their moisture. As a result, the sheltered leeward slopes receive little rain. Aridity is further increased because air descending from the mountains is warmed by compression. Extreme examples of the influence of topography on aridity include:

Figure 6.3 Rain shadow effect

- the Kalahari Desert in southern Africa, which lies in the rain shadow of the Drakensberg mountains
- the Atacama Desert in Chile and Peru, in the rain shadow of the Andes (Figure 6.3)

2.3 Cold ocean currents

Cold ocean currents flow along the western margins of continents in the tropics and sub-tropics. They are directly responsible for the extreme aridity of coastal deserts in Namibia and northern Chile/southern Peru. In Namibia, the prevailing southeast **trade winds** push coastal surface waters offshore (Figure 6.4). This allows the upwelling of cold water from depth and the formation of the cold Benguela surface current. Air in contact with the ocean surface is chilled and forms a **temperature inversion**.

The inversion is spread onshore by local winds and prevents convection and cloud formation. This process, combined with the already dry air caused by the rain shadow effect of the Drakensberg mountains and the descent of air to the coast from the African plateau, results in one of the driest places on the planet: the Namib Desert.

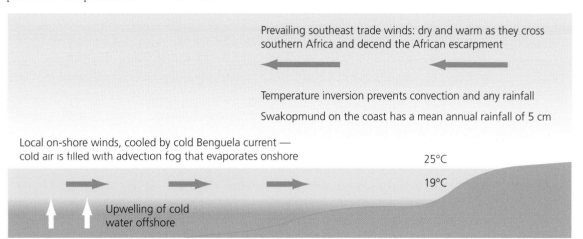

Figure 6.4 Southeast trade winds in the Namib Desert

3 Weathering processes

The main weathering processes in hot arid and semi-arid environments are salt weathering, insolation weathering and freeze–thaw. These processes are described in detail in Topic 3. Their effectiveness can be seen in the large amounts of rock debris found in hot deserts.

Most types of weathering require moisture. Even in hyper-arid areas it rains occasionally and moisture from dew is common at night. Meanwhile, in more humid, semi-arid environments, there is usually a wet season lasting for 2 or 3 months when moisture is available.

Freeze–thaw can be surprisingly active in sub-tropical desert environments, particularly where altitudes exceed 1,500 m. Clear skies and the relatively thin atmosphere at altitude cause significant heat loss at night in winter. Frost is a common occurrence in the Kalahari, Mojave and northern Sahara deserts.

4 Aeolian processes and landforms

Aeolian processes describe the transportational, erosional and depositional action of the wind. The sparse vegetation cover and dry conditions in hot arid and semi-arid environments makes wind action more effective than in humid environments.

4.1 Aeolian transport processes

The wind transports sand and silt particles in three ways: by creep, saltation and suspension.

Creep occurs when sand grains slide and roll across the surface. It is caused by drag and small differences in pressure that create lift.

Saltation is the downwind skipping motion of sand grains. It occurs only within 1–2 m of the surface. When saltating grains hit a sand surface, they have a ballistic effect and set other grains moving in the direction of the wind.

Small dust particles (less than 2 mm in diameter) can be entrained by the wind and transported in **suspension** beyond desert areas. In north Africa large amounts of wind-blown dust are deposited in the Atlantic Ocean every year. Deposits of wind-blown dust, transported at the end of the last glacial, are common in Europe and China and are known as **löess**.

4.2 Aeolian erosion and landforms

The main erosional effect of the wind is the removal of fine particles in a process called **deflation**. Silt and clay-sized particles entrained by the wind can be transported thousands of kilometres. Locally, deflation is responsible for dust storms and for surface lowering, which forms shallow depressions. The selective erosion and transport of fine material by deflation often leaves behind coarse lag particles. The result is extensive surfaces dominated by coarse rock particles known as **desert pavement** (Figure 6.5). Stony surfaces formed in this way are known as **reg** in the Sahara.

The scouring effect of sand-sized particles saltated by the wind is known as **aeolian abrasion**. Although abrasion may polish rock surfaces, its overall importance is small. Because sand grains are relatively heavy, the effect of wind abrasion is confined to within 1–2 m of the surface. Faceted cobbles and pebbles (**ventifacts**) are small-scale erosional features formed by wind action.

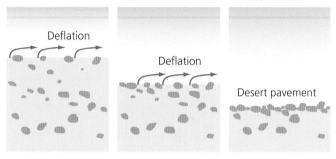

Figure 6.5 Desert pavement

Yardangs are the most impressive landforms caused by aeolian erosion. They are streamlined ridges of solid rock, aligned in the direction of the prevailing wind. In scale they vary from 10 m to 100 m in height and are often undercut at the base — evidence that abrasion by saltating sand grains is instrumental in their formation.

4.3 Aeolian deposition and landforms

Sand dunes are the main landform of aeolian deposition. In hot arid and semi-arid environments, sand accumulates in vast sheets or 'sand seas', known as **ergs** in the Sahara. Sand accumulation occurs in areas of reduced wind speed. Once sand is deposited, it attracts further deposition because saltating sand grains are less able to rebound off a soft sand surface compared to a rocky surface. As a result, areas of dune are often fairly localised and surrounded by alluvial plains and rocky deserts.

Dunes are accumulations of blown sand that form mounds and ridges. Two conditions are needed for dune formation: an adequate supply of sand, and winds strong and persistent enough to transport the sand.

A typical dune formed by winds from a prevailing direction has a windward slope of 10–15°, a sharp crest and a much steeper leeward or **slip face** of 30–35° (Figure 6.6). The slip face stands at the angle of repose, i.e. the maximum angle at which loose sand is stable. Creep and saltation transport sand up the windward slope. As sand accumulates on the crest, it eventually exceeds the angle of repose, causing small avalanches down the slip face which restore equilibrium. In this way dunes advance in the direction of the prevailing wind.

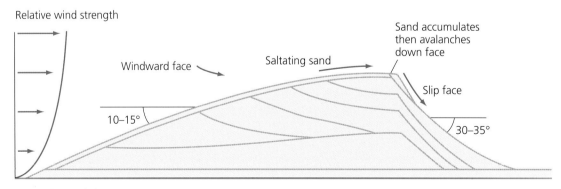

Figure 6.6 Sand dune

Types of dune

Dunes are classified into four main types according to their shape in planform (Figure 6.7). Several factors influence dune types, including sand supply, wind direction and vegetation cover.

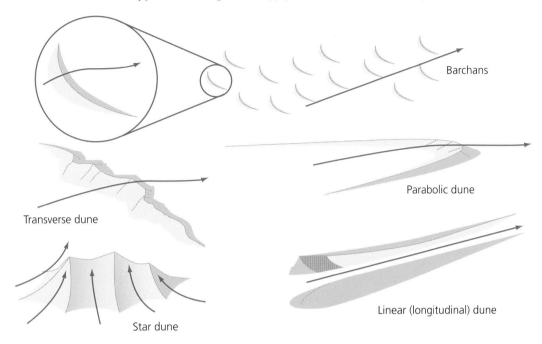

Figure 6.7 Types of sand dune

Crescentic dunes

Crescentic dunes are wider than they are long, with slip faces on their concave side (in plan). The two main types of crescentic dune are **barchans** and **transverse dunes**. Barchans are easily recognised by horns that face downwind and, like transverse dunes, form where winds blow predominantly from one direction. Crescentic dunes are highly mobile, moving up to 100 m in a year.

Linear dunes

Linear dunes are straight or slightly curved in planform. They are often more than 100 km long with slip faces on alternate sides. They occur either as isolated ridges or as a series of parallel dunes and cover a larger area of desert than any other dune type.

Star dunes

Star dunes are pyramidal with slip faces on three or more arms that radiate from a dome-line summit. They form in areas where the wind is multi-directional.

Parabolic dunes

Parabolic dunes have U-shaped planforms with convex noses trailed by elongated arms. Unlike crescentic dunes, the arms of parabolic dunes extend upwind. The arms are fixed by vegetation while the main mass of the dune moves forward.

5   *Fluvial processes and landforms*

Many characteristic desert landforms owe their formation to streams and rivers. These features often have their origin in the past, when hot arid and semi-arid environments supported a more humid climate that at present.

Even though most desert streams and rivers are ephemeral and flow for only part of the year, some fluvial landforms are still developing today. Dry river channels, known as **wadis** in Africa and arroyos in North and Central America, are common and often occur at surprisingly high densities. This is because rates of runoff relative to rainfall are high due to:

- sparse vegetation cover, with minimal interception and transpiration and extensive bare rock surfaces
- shallow soils, which often comprise no more than a veneer of gravels and sands and provide little water storage
- the ground surface, which is often baked hard and limits infiltration
- rainsplash on unvegetated surfaces, which fills soil pores and reduces soil permeability

On steep slopes made of softer rock like shale and marl, runoff erodes parallel V-shaped gulleys and channels separated by narrow, knife-edged ridges. Over wide areas this produces a **badlands** type of topography.

Thunderstorms and intense convectional downpours are typical of hot arid and semi-arid environments. For short periods, streams and rivers sustain high discharges. At these times they are effective agents of erosion and transport large volumes of weathered rock debris. The result is landforms such as alluvial fans, bajadas, playas and canyons.

5.1 Alluvial fans and bajadas

Alluvial fans are cones of debris that accumulate along mountain fronts where a river leaves a steep-walled valley (canyon) in the mountains and enters an adjacent basin or lowland. On entering the basin, the reduction in gradient causes a sudden loss of energy and the deposition of alluvial sediment (see Figure 3.7 on page 25). The main channel splits into hundreds of smaller channels, which results in a delta-shaped mound of debris with a shallow gradient and concave profile. Where multiple alluvial fans develop, they often merge to form a continuous apron of debris known as a **bajada**.

5.2 Playas

Most streams and rivers in hot arid and semi-arid environments drain to shallow inland basins, where they form temporary lakes or **playas**. The lakes soon evaporate, leaving behind salts such as sodium chloride, sodium sulphate and gypsum. In more humid environments these salts would normally be transported to the sea. In deserts they accumulate and may form extensive salt flats like those at Bonneville in Utah, USA.

5.3 Canyons

Canyons are narrow river valleys with near-vertical sides cut into solid rock. They are common in mountains and plateaux in hot arid and semi-arid environments. Canyons result from vertical erosion by rivers and streams. Erosion occurs vertically rather than laterally because solid rock walls allow little lateral migration of river channels. Also important in canyon development is the absence of slope processes found in more humid climates such as soil creep, slumping and mudflows. These processes normally backwaste slopes, widening out valleys. In regions such as the Colorado Plateau in the southwestern USA and the Anti-Atlas mountains in North Africa, **tectonic** uplift has assisted canyon incision.

6 Human impacts in hot arid and semi-arid environments

Hot arid and semi-arid environments are fragile and easily degraded by human activity. Ecological fragility is due to:
- low rates of primary production and plant growth caused by sparse rainfall and high temperatures
- low biodiversity. With few species adapted to the harsh environments of deserts, the loss of a single species can destabilise ecosystems and cause severe disruption

The implication of fragility is that hot arid and semi-arid environments are slow to recover from the effects of **overcultivation**, **overgrazing** and **deforestation**. Once the vegetation cover is damaged, the result is often widespread soil erosion and land degradation.

6.1 Desertification

Desertification describes the degradation of formerly productive land to the point where desert-like conditions prevail. There is a gradual loss of biological and economic activity as water, soil and vegetation resources are degraded (Figure 6.8). One-third of the Earth's land surface is threatened by desertification, directly affecting 250 million people (Figure 6.9). Annually, an estimated 1.5 to 2.5 million ha of irrigated land and 3.5 to 4.0 million ha of red-fed agricultural land lose all or part of their productivity because of land degradation.

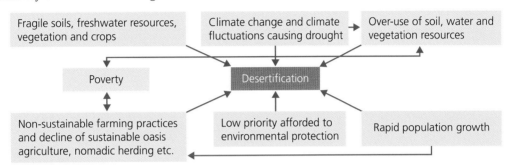

Figure 6.8 Desertification

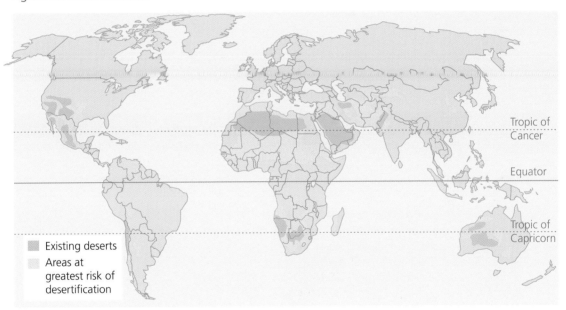

Figure 6.9 Areas at risk of desertification

6.2 Land degradation

Land degradation takes one or more of three forms in hot arid and semi-arid environments:

- soil erosion
- salinisation
- deforestation

Land degradation reduces the productive potential of farmland and impoverishes the lives of millions of people. In extreme cases it leads to food shortages and permanent damage to soil, water and vegetation resources.

Soil erosion

Soil erosion is the accelerated loss of soil by wind and surface runoff. In hot arid and semi-arid environments, the main triggers of soil erosion are:

- deforestation caused by overgrazing or deliberate clearance for fuelwood and timber
- overcultivation, which depletes the soil's organic material and destroys its structure

Soil erosion also results in secondary problems, including damage to crops by blown soil, slopes carved by runoff into deep gullies and the siltation of irrigation canals and reservoirs.

Salinisation

Salinisation is the accumulation in soils of salts that are toxic to plants, often to the point where agriculture is abandoned. Most salinisation of farmland is due to **overirrigation** and inadequate drainage. The outcome is a rise in the water table until salty water reaches the root zone of crops. In extreme cases water is drawn to the surface by capillary action where it evaporates, leaving behind a salt crust (Figure 6.10).

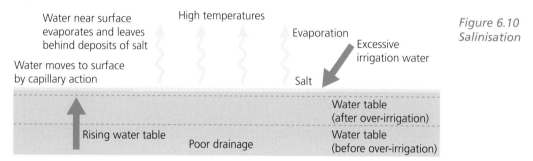

Figure 6.10 Salinisation

Deforestation

Pressures on woodland are acute in many regions in hot arid and semi-arid environments. The demand for firewood for cooking and lighting has led to extensive desertification in the African Sahel, in countries like Burkina Faso, Mali and Niger. In some areas virtually all trees have been removed for several kilometres around villages. Meanwhile, overgrazing by livestock prevents any regeneration. **Deforestation** exposes fragile soils to erosion by wind and water.

CASE STUDY 1	**Land degradation in the Sahel: Nara, Mali**
Background	Nara is a poor region in the impoverished state of Mali in West Africa (Figure 6.11). It is a tropical semi-arid region in the Sahel, on the southern fringes of the Sahara Desert. Mean annual rainfall is 378 mm; potential evapotranspiration is four to five times greater. Rainfall has a seasonal pattern. There is a wet season between June and October; the rest of the year is dry. Drought is common, and plants, animals and indigenous people are adapted to drought. Indigenous people comprise: • agro-pastoralists (e.g. Sarakolé), who are sedentary • nomadic pastoralists (e.g. Fulani)
Desertification	Desertification began with the severe drought that hit the Sahel between 1969 and 1974. Since then, rainfall has fallen by 30%, but the human population has grown rapidly. As a result, demand for food, fodder and firewood have risen steeply. Drought and population growth have placed excessive pressures on environmental resources, leading to deforestation, overgrazing and overcultivation. The outcome is desertification and land degradation, which has lowered yields, caused food shortages and exacerbated poverty. Many indigenous people have been forced off the land and have migrated to the capital city, Bamako.

Figure 6.11 Location of Nara, Mali

Other conse-quences of desertification	There is no tradition of managing pastureland in Nara. Consequently, overgrazing has occurred and pasture quality has declined. In response to population pressure, agro-pastoralists have extended cultivation into marginal areas at the expense of woodland. Shortages of pasture have led to conflicts between nomadic groups and agro-pastoralists. Drought has increased the salt content of groundwater and overgrazing has reduced biodiversity. Wildlife such as rabbits and antelope, once an important food source for local people, has largely disappeared, and years of plentiful rain (2003, 2004) triggered plagues of locusts that destroyed crops.
Conclusion	Nara occupies the Sahel region, which stretches across northern Africa from Senegal to Eritrea. This is a fragile environment where ecosystems and human activities are finely balanced. A combination of drought and population growth has led to the unsustainable exploitation of woodland, soil and water resources. The result has been land degradation and desertification with falling water tables, soil salinisation, deforestation, overgrazing and ultimately soil erosion. As the natural wealth of Nara has degraded, food shortages, societal breakdown and out-migration have occurred.

7 Managing desertified and degraded land

Reversing the effects of desertification and land degradation is not easy. Fragile hot arid and semi-arid environments recover slowly from the effects of deforestation, overgrazing and salinisation. Responses include:
- reafforestation programmes
- construction of shelterbelts to combat wind erosion
- land drainage to rehabilitate salinised farmland
- destocking to reduce the pressure of grazing on open rangeland
- educating farmers in techniques of sustainable land management

CASE STUDY 2 The Korqin sandy lands and China's Great Green Wall

Background	Environmental degradation leading to desertification affects one-third of China's total land area. The most seriously affected regions are the arid and semi-arid regions of northern China. China loses 5 billion tonnes of topsoil to erosion every year. The Korqin sandy lands of northern China are highly susceptible to wind erosion and include some of the most degraded land in China (Figure 6.12). Land degradation and desertification are the result of overgrazing and clearing trees for timber and farmland, which in turn are related to population pressure. Overexploitation of groundwater has lowered the water table over large areas and poor drainage has caused salinity problems. Strong winds are the main agents of desertification and produce violent dust storms.
Management responses	In 1978 China embarked on an ambitious programme to combat desertification in its northern provinces. The scheme, popularly known as the Great Green Wall, is a massive reafforestation programme. The aim is to establish 350,000 km² of plantation forests and shelterbelts across the entire region by 2050. So far, 130,000 km² have been planted. In Korqin the programme will protect farmland against erosion, restore soil fertility and improve the well-being of local inhabitants. It will also provide a sustainable source of timber. Plantations and shelterbelts have been established using native poplar trees, which are resistant to drought and frost. However, the scheme is not just about reafforestation. There is an emphasis on conservation, with appropriate cultivation techniques involving: • the recycling of organic material to the soil • integrating tree crops with pasture and cash crops (agro-forestry) • planting tree species that provide fodder and timber and improve soil fertility Meanwhile, the controlled management of grazing lands (i.e. keeping within the land's carrying capacity) is being introduced for the first time.

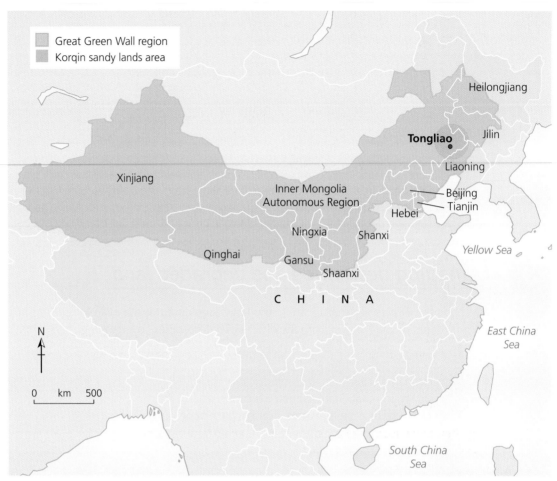

Figure 6.12 The Korqin sandy lands, China

TOPIC 7 Climate change and climate hazards

1 Climate change

Long-term natural climate shifts are caused by astronomical events such as changes in the tilt of the Earth's axis and last for thousands of years. Medium-term changes, lasting a few hundred years, result from the shutdown of warm ocean currents. Short-term climatic change, which may last for several years, is often triggered by major volcanic eruptions.

1.1 Long-term change: glacial cycles

Over geologic time the Earth's climate has undergone many natural changes. Most dramatic have been the **glacial cycles**. Glacial periods (ice ages), when ice covered around one-third of the land surface, have dominated the past 1.5 million years (the Pleistocene period). Separating the cold glacials were warmer **interglacial** periods. Currently the world is in one of these warmer phases (the Holocene), which has lasted for 10,000 years.

The main causes of glacials are changes to the Earth's axis and orbit, which affect the amount of solar radiation reaching the planet's surface.
- Today, the Earth is tilted on its rotational axis at an angle of 23.4° to its orbital plane. However, over a period of 41,000 years the angle of inclination fluctuates between 22 and 24.5°.
- The eccentricity of this Earth's orbit around the sun varies from near circular to markedly elliptical. The two extremes are separated by 96,000 years.
- The Earth gyrates like a spinning top on its axis. This phenomenon, known as the precession of the equinoxes, affects the intensity of the seasons.

Together, these astronomical changes combine to create cycles of glacials and interglacials. They are known as **Milankovitch cycles**.

1.2 Medium-term change: ocean currents

Shifts in surface ocean currents can cause sudden climate change. The warm North Atlantic current shut down between 11,000 and 10,000 years ago and brought ice age conditions to northwest Europe (the so-called **Loch Lomond stadial**). The resulting cooldown lasted for nearly a millennium, causing glaciers to re-form in Scotland, the Lake District and Wales. Scientists think that **global warming** could trigger a similar cooldown some time in the twenty-first century.

1.3 Short-terms change: volcanic eruptions

Major volcanic eruptions can cool the global climate for several years and may have triggered glaciations in the past. Large-scale eruptions pump huge amounts of volcanic ash into the stratosphere, reducing **insolation** at the Earth's surface. In addition, sulphur dioxide from eruptions persists as tiny droplets of sulphuric acid, which reflect insolation. Mt Pinatubo, which erupted in June 1991, cooled the global climate for a couple of years, overriding global warming.

2 Evidence for climate change

Evidence for climate change comes from several sources. These include sea-floor sediments, ice cores, pollen analysis and dendrochronology. The forensic evidence is described in Table 7.1.

Table 7.1 Forensic evidence for climate change

Source	Description
Sea-floor sediments	The fossil shells of tiny sea creatures called foraminifera, which accumulate in sea-floor sediments, can be used to reconstruct past climates. The chemical composition of foraminifera shells indicates the ocean temperatures in which they formed.

Ice cores	Ice cores from the polar regions contain tiny bubbles of air — records of the composition of the atmosphere in the past. Scientists can measure the relative frequency of hydrogen and oxygen atoms with stable isotopes. The colder the climate, the lower the frequency of these isotopes.
Pollen analysis	Pollen analysis allows scientists to reconstruct past vegetational changes. From these data we infer palaeoclimatic conditions. Pollen diagrams show the number of identified pollen types in the different sediment layers as a direct count or as a percentage of total tree pollen (Figure 7.1).
Dendrochronology	Dendrochronology is the dating of past events such as climate change through study of tree ring (annule) growth. Annules vary in width each year depending on temperature and moisture availability.

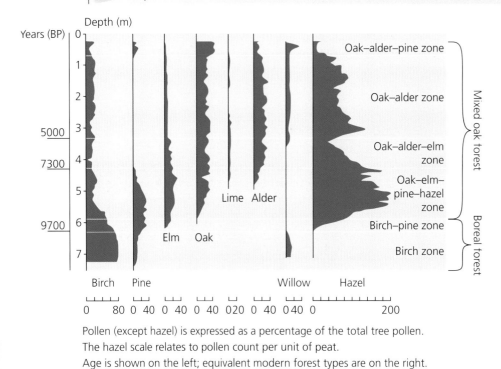

Pollen (except hazel) is expressed as a percentage of the total tree pollen.
The hazel scale relates to pollen count per unit of peat.
Age is shown on the left; equivalent modern forest types are on the right.

Figure 7.1 Tree pollen counts

2.1 Historical records

Historical records also provide information about past climates. Europe experienced a so-called Little Ice Age between the mid-fourteenth and early nineteenth centuries. The seventeenth-century diarist John Evelyn described frost fairs on the River Thames in London, when the river froze over. Such events were captured by contemporary artists like Abraham Hondius. The sixteenth-century Dutch artist Pieter Breugel the Elder is famous for his winter landscapes, with snow, frozen lakes and skaters. In 1783 many scientists and naturalists (including Gilbert White and Benjamin Franklin) recorded the effects of the eruption of the Laki volcano in Iceland in Europe and North America. They described ashfalls and a thick sulphurous smog that settled over much of the northern hemisphere. They also noted the cooling effect of the eruption on the summer of 1783 and the exceptionally severe winter that followed.

3 Global warming: anthropogenic climate change

There is conclusive evidence that the Earth's climate has warmed in the past 50 years (Figure 7.2). Globally, ten of warmest years on record occurred between 1997 and 2008. Average global temperatures for 2000–08 were 0.2°C higher than during the previous decade. The highest-ever temperature in the UK (38.5°C) occurred in August 2003 and the highest July temperature (36.1°C) was recorded in 2006 (Figure 7.3).

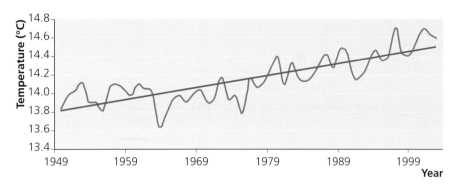

Figure 7.2 Global temperature change, 1950–2004

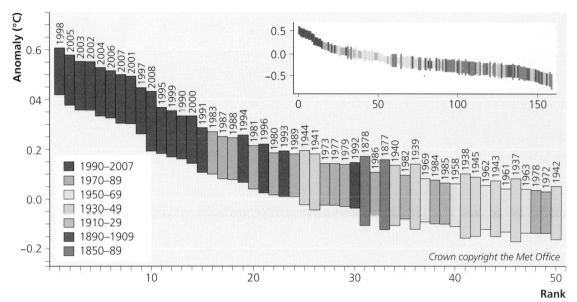

Figure 7.3 Global warming, 1850–2007

Today, the overwhelming view of scientists is that global warming is a reality, but there is still debate on its cause. Is global warming a natural trend or is it due to human activities? We know that the global climate undergoes periodic change due to astronomical cycles, switches in surface ocean currents and volcanic eruptions. However, current climate change is different because it is happening much faster than any previous event and there is overwhelming evidence linking it to human activities. Climatologists are in broad agreement that average global temperature (14.3°C in 2008) is 0.7°C higher in recent years because of human activities.

> ← Humans did it

3.1 Anthropogenic warming: the greenhouse effect

There is a strong correlation between the rise in average global temperatures during the past 60 years and the volume of carbon dioxide in the atmosphere (Figure 7.4). Before 1800, average carbon dioxide concentrations were around 270 ppm. Today's concentrations are an average of 387 ppm and rising rapidly. This is due largely to the burning of fossil fuels, although deforestation and draining wetlands have also played a part.

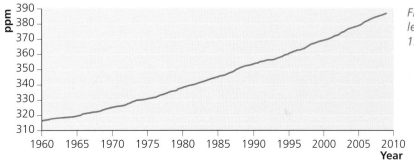

Figure 7.4 Carbon dioxide levels in Manua Loa, Hawaii, 1960–2009

The greenhouse effect explains the link between global temperatures and carbon dioxide levels. **Greenhouse gases** (GHGs) such as water vapour, carbon dioxide and methane occur naturally in the atmosphere. They absorb and re-radiate around 95% of long-wave radiation emitted by the Earth. Carbon dioxide alone raises average global temperatures by 7°C. Without this natural greenhouse effect average global temperatures would be 30°C lower. However, large increases in carbon dioxide and other GHGs during the past two centuries have led to more absorption of long-wave radiation by the atmosphere. The result is an enhanced greenhouse effect inducing global warming and climate change.

bring about, enhance

4   The impacts of global warming

4.1 Climate change

The most direct effect of global warming is climate change, which will affect billions of people. Computer models predict major disruption to regional rainfall later in the twenty-first century. Although some regions will get wetter, large parts of North America, South America, southern Europe, Africa, the Middle East and central Asia will experience lower rainfall and more frequent droughts.

Drought in Portugal and Spain in 2004 and 2005 has been linked to global warming. The hot, dry summer of 2005 followed an exceptionally dry winter. In Portugal 97% of the country was hit by severe drought — the worst since 1940. Wildfires burned out of control and destroyed an estimated 2,400 km² of forest and farmland.

The Intergovernmental Panel on Climate Change (IPCC) predicts an average rise in global temperatures of around 3°C by the end of the twenty-first century. However, continued growth in the use of fossil fuels could see global temperatures rising by as much as 5°C. Levels of warming will increase from the equator to the poles. Temperatures in the Arctic and sub-Arctic in winter could rise by 10°C to 18°C.

In mid-latitudes severe storms will be more frequent, increasing coastal erosion and coastal flooding. Meanwhile, warmer conditions will intensify the water cycle causing more evaporation, intense rainstorms and river flooding. In the tropics and sub-tropics, warmer ocean waters will generate more powerful tropical storms and hurricanes.

change in volume of sea due to a change in temperature

4.2 Rising sea level

Melting glaciers and ice sheets, together with the **thermal expansion** of the oceans, are responsible for rising sea levels — levels rose by nearly 20 cm between 1900 and 2000. Computer models forecast an average rise in sea level of 40 cm by the end of the twenty-first century (Figure 7.5). This could spell disaster for Bangladesh, where 37% of the country is less than 3 m above sea level. Even worse, within the next 50 years island states like the Maldives and Tuvalu could disappear altogether.

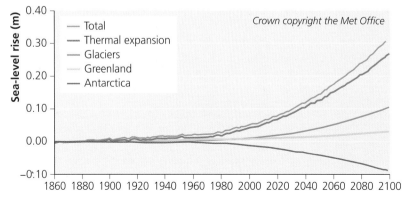

Figure 7.5 Model showing sea level change

In the UK rising sea levels will lead to more coastal flooding. If London's sea defences were breached, it could cause damage up to £25 billion. Sea level rise and stormier conditions also threaten the UK's coastal defences. In future the costs of maintaining sea walls and other coastal defence structures will be unsustainable. Along some coastlines sea defences will be abandoned and nature left to take its course. Elsewhere, defences will be dismantled as part of a policy of **managed realignment**. Either way, large tracts of reclaimed farmland along the coast will revert to salt marsh and mudflat.

4.3 Water resources and farming

Some 16% of the world's population and one-quarter of global economic output could be hit by water shortages caused by climate change. With 98% of the world's glaciers currently retreating, areas that rely on meltwater are most at risk.

Computer forecasts for the 2080s suggest that global warming will reduce summer rainfall by up to 40% in southern Britain. Water resources would be further reduced by a possible 3°C temperature rise, increasing evaporation and reducing runoff to half current levels.

Drylands in southern Europe and North Africa are already marginal for farming. Any significant decline in rainfall could lead to land degradation and desertification. Other regions likely to be affected by drier conditions include commercially important farming areas like the Prairies in the USA and Canada, and the Pampas in Argentina. As the climate dries out, cereal production will slump. Some experts believe that production could drop by as much as 400 million tonnes a year, leading to global food shortages.

4.4 Environmental impacts

Climate change puts huge pressure on natural **ecosystems**. Habitats will change and species will have to adapt by migrating either latitudinally or altitudinally. Species that cannot respond in this way (e.g. those already occupying extreme environments such as the Arctic and sub-Arctic) will face extinction.

CASE STUDY 1	The impact of climate change on the North American Arctic tundra
Natural ecosystem	Occupies northern Canada and much of Alaska. Dominated by low-growing, woody plants (e.g. heather, mountain avens), grasses, mosses and lichens. Only dwarf tree species can survive. Sub-zero temperatures for 9 months per year and a short growing season. Permafrost, with lakes and marshes during summer when surface layers melt. Low plant productivity and low biodiversity. Short food chains. Both plant and animal species are highly specialised. Large migrations of wildfowl and wading birds in summer take advantage of the glut of insects. Migrations of caribou to the tundra in summer for calving and grazing, followed by predators (e.g. wolves, wolverines, arctic foxes). Other herbivores include lemmings, arctic hares and musk oxen. The tundra is a fragile environment with little resilience to adapt to climate change. Arctic marine ecosystems include seals, walruses, polar bears, killer whales and narwhals. Sea ice plays a crucial role in the marine ecosystem.
Climate change	The tundra is cold and relatively dry. Global warming will bring profound changes. Large temperature rises are forecast for Arctic and sub-Arctic environments: 5 or 6°C by the end of the century. This is primarily due to decreases in **albedo** (surface reflectivity) as snow, ice and sea ice melt, which leads to increased absorption of solar radiation (Figure 7.6). The poleward advance of the tree line will also lower albedos and accentuate warming. With higher temperatures, precipitation will increase.
Biotic change	Many tundra species will be intolerant to warmer conditions and higher rainfall but will have nowhere to go. Trees will invade as the treeline advances northwards, changing the open tundra to boreal forest. Existing habitats will shrink and become fragmented. About 90% of Alaska's tundra will disappear by the end of the century. Habitat change will have drastic effects on food webs and wildlife populations and many species will not survive. Melting of sea ice will prevent seals and walruses hauling out on the surface. Polar bears live on the sea ice while they hunt for seals and other marine mammals. As the ice disappears, this will become impossible. Warmer conditions will see more insects and pathogens, spreading plant diseases. As the forest advances, migrant birds will lose their summer wetland breeding habitats.

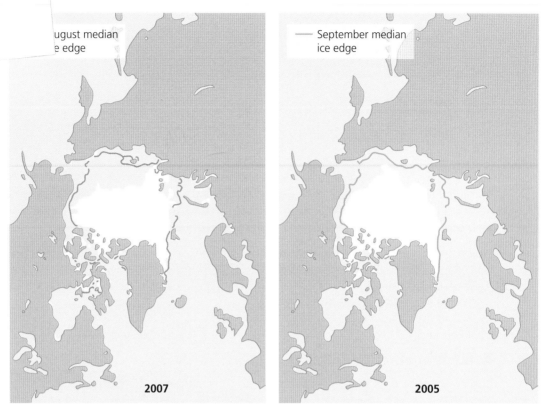

Figure 7.6 Arctic sea ice, 2007 and 2005

Source: NSIDC

5 Predicting climate change

Complex computer models are used to predict climate change. These predictions are based on a range of projected carbon dioxide emissions and levels of economic growth. Figure 7.7 shows the predictions from the Met Office's climate model for annual average temperature by the end of the twenty-first century. Even if emissions of carbon dioxide and other GHGs are drastically reduced, because of the long residence time of carbon dioxide in the atmosphere climate change is already inevitable.

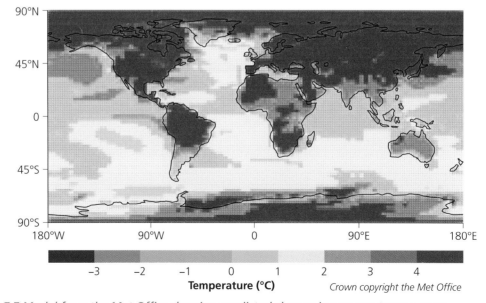

Figure 7.7 Model from the Met Office showing predicted change in average temperature

6 *Responding to climate change*

Responses to climate change occur at international, national and individual scales.

6.1 International

Climate change is a transnational environmental problem. Tackling problems at this scale requires international cooperation and agreements that are often hard to achieve. So far, the only truly international initiative on global climate change is the **Kyoto Protocol**. This treaty dates back to 1997, although it only became international law in 2005. Countries have agreed to limit emissions of carbon dioxide to achieve a 5.2% reduction on 1990 outputs by 2012. Although it is an important first step in the drive to tackle climate change, Kyoto has had only limited success. There are several reasons for this:

● Not all countries have ratified the treaty. The USA, the world's largest producer of carbon dioxide, has not signed up.
● Only 37 of the 183 countries that have ratified Kyoto have targets for a reduction in emissions: developing countries are exempt. Therefore, major polluters such as China, India and Brazil do not have to limit their emissions (Figure 7.8).
● Most of the 37 countries will fail to achieve their 2012 targets.
● Rapid economic growth in emerging economies (often based on coal) will nullify any gain from emissions reductions in developed countries.
● No agreement has yet been reached on what will replace Kyoto when it expires in 2012.

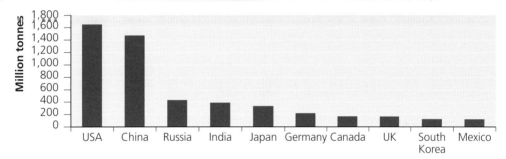

Figure 7.8 Carbon dioxide emissions

An alternative approach at international level is **carbon trading**. Under this scheme, businesses are allocated a quota for carbon dioxide emissions. If they emit less than their quota they receive carbon credits, which can be traded on international markets. Companies that exceed their quota must purchase additional allowances or pay a financial penalty.

6.2 National

It is easier for individual governments to develop their own approaches to control **carbon emissions**. Governments may:

● give subsidies to promote renewable energy such as wind and solar power, as well as nuclear power
● impose a carbon tax on activities that use large amounts of fossil fuel
● encourage investment in new and/or cleaner technologies

6.3 Individual

Individuals can contribute to lower carbon emissions by reducing their **carbon footprint**. Examples include energy conservation through better home insulation, walking or cycling to work, purchasing locally grown food, using recycling schemes and buying hybrid cars or more fuel-efficient vehicles. **Carbon offsets** encourage individuals (and companies) to take responsibility for their carbon emissions. Purchasing carbon offsets compensates for the emissions by funding an equivalent carbon dioxide saving elsewhere.

7 Climatic hazards

Extreme weather events such as hurricanes, tornadoes, depressions and droughts often result in loss of life and injury. They also damage and/or destroy property, infrastructure and livelihoods. When climatic events have an adverse effect on economy and society, they become **natural hazards**.

8 Hurricanes — UNIT 2

Violent tropical storms in the Atlantic region are known as **hurricanes**. Similar storms in east Asia are called **typhoons** and in south Asia they are known as **tropical cyclones**. These storms have a number of defining features:

● sustained wind speeds of more than 119km/h
● exceptionally heavy rainfall
● deadly storm surges in coastal areas

8.1 Formation and decay

Hurricanes form over tropical oceans between latitudes of 8 and 20°N. Three conditions favour hurricane formation:

● high humidity and therefore plenty of water vapour
● light winds, which allow significant vertical cloud development (i.e. little or no wind shear)
● sea surface temperatures (SSTs) of at least 26–27°C

These conditions occur in summer and early autumn in the tropical North Atlantic and North Pacific oceans. As a result, the hurricane season in the northern hemisphere runs from June to late October. By November ocean waters are too cool to generate hurricanes.

The first signs of a hurricane are **tropical disturbances**: clusters of thunderstorms that develop over the ocean. Given favourable conditions, some of these disturbances become better organised. As pressure falls in the area around the storm and condensation releases latent heat, warm air rises. Soon, a distinctive cyclonic circulation (anticlockwise in the northern hemisphere) develops in response to the Earth's rotation. Meanwhile, rising air from evaporation cools, condenses and releases more latent heat, creating low pressure at the surface and high pressure near the top of the storm (Figure 7.9).

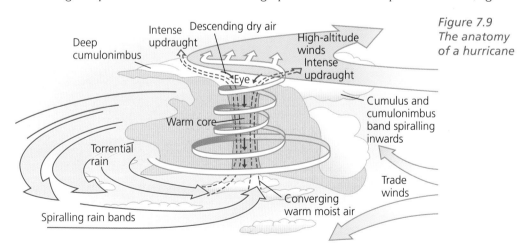

*Figure 7.9
The anatomy
of a hurricane*

As a result, the central area of a hurricane behaves like a giant chimney, with high pressure aloft forcing air outwards and low pressure at the surface sucking air in. In this way the hurricane gets a constant supply of fresh vapour — the energy that drives the storm. Once sustained wind speeds reach 37km/h, tropical disturbances become **tropical depressions**. As winds increase to 63km/h, tropical depressions are upgraded to **tropical storms**. Eventually, when wind speeds reach 119km/h, tropical storms achieve hurricane status (Table 7.2).

Table 7.2 Hazards associated with hurricanes

Feature	Description
Damaging winds	Winds can reach 250 km/h in the wall of the hurricane, with gusts of up to 360 km/h. Destruction results from both direct impact and flying debris. Violent winds damage trees and crops. Tall buildings can collapse. Sudden pressure changes may cause buildings to explode and suction can lift roofs and entire buildings. However, most destruction, death and injury is due to flying debris.
Storm surges	A storm surge is water pushed towards the shore by the force of the winds. An advancing surge combined with high tides can increase mean water level 5 m or more. Meanwhile, low atmospheric pressure pushes up the sea surface. The resulting rise in water level can cause severe flooding in coastal areas. Storm surge heights are also determined by the slope of the continental shelf. A shallow offshore gradient allows a greater surge.
Heavy rainfall	Rainfall is often heavy and prolonged. Intense rainfall causes three types of hazard: • water seepage into buildings, which may result in their collapse from the weight of the absorbed water • inland flooding by rivers • mass movements such as landslides, mudslides, mudflows and debris flows

Although hurricanes take weeks to form, they often disappear in just a few days. Rapid decay takes place when a storm moves:

- over cooler water that can no longer supply enough vapour
- over land, where it abruptly loses its power source of the warm, surface waters of the ocean
- into an area where the large-scale flow aloft is either subsiding or where wind shear is strong

Box 1   *The Saffir-Simpson scale*

The Saffir-Simpson scale grades hurricanes on a 1 (weakest) to 5 (strongest) scale. The scale takes account of a hurricane's central pressure, maximum sustained winds and storm surge (Table 7.3). Sustained wind speeds are the defining factor, as storm surge values are highly dependent on other factors such as the slope of the continental shelf in the landfall region.

Categories 3, 4, and 5 are major hurricanes, capable of inflicting great damage and loss of life.

Table 7.3 The Saffir-Simpson scale

Scale number	Central pressure (mb)	Wind speed (km/h)	Storm surge (m)	Damage
1	> 980	119–153	1.2–1.5	Minimal
2	965–979	154–177	1.6–2.4	Moderate
3	945–964	178–209	2.5–3.7	Extensive
4	920–944	210–250	3.8–5.5	Extreme
5	< 920	> 250	> 5.5	Catastrophic

8.2 Managing hurricanes

Hurricanes are more closely monitored than any other natural hazard. Sophisticated measurement of temperatures, humidity and wind speed, and tracking storm paths using satellites, aircraft, ships and buoys, means that accurate forecasts can be issued by agencies such as the US National Weather Service. The US National Hurricane Center measures and monitors hurricanes in the Atlantic and eastern Pacific.

Storm surges are the most deadly hazard caused by hurricanes. On the Gulf coast in the USA, major cities are protected by flood embankments or levées. In developing countries such as Bangladesh and

India, storm shelters built on stilts provide temporary refuge from storm surges. Where storms are closely monitored, early warning may trigger mass evacuation, e.g. New Orleans in 2005 (Hurricane Katrina), Bangladesh in 2007 (Cyclone Sidr) and Houston in 2008 (Hurricane Ike).

Box 2   *Measuring and monitoring hurricanes*

Monitoring begins in the early stages of storm development in the ocean using satellites, ships and buoys moored at sea. Closer to land, direct measurements are made by aircraft and radiosondes. Data are fed into computer models which forecast storm intensities and tracks.

Geostationary satellites provide information on the size, intensity and movement of storms. **Ships and buoys** provide air temperature, sea surface temperature, wind speed, wind direction, pressure and humidity data. **Aircraft** fly into the storms to measure wind speed, pressure, temperature and humidity.

Radiosondes are small instrument packages and radio transmitters that are attached to balloons and released into storms. They provide additional information on temperature, wind speed, pressure and humidity. **Radar** images provide information on rainfall intensity.

The National Weather Service issues two categories of warning of approaching hurricanes:
- **Hurricane Watch**: announcements for specific coastal areas that hurricane conditions are possible within the next 36 hours
- **Hurricane Warning**: announcements for specific coastal areas that sustained winds of 119 km/h and above are expected within the next 24 hours

CASE STUDY 2	Hurricane Katrina, 2005
Physical details	Hurricane Katrina hit the Gulf Coast of Louisiana on 29 August 2005. The storm killed 1,422 people, of whom 1,104 were in Louisiana. Some 350,000 people were evacuated. The damage ($75 billion) made Katrina the costliest natural disaster in US history. The storm surge that flooded 80% of New Orleans was responsible for most of the deaths (Figure 7.10).
Exposure	Hurricane Katrina was a category 5 storm and the fourth most powerful hurricane to make landfall in the USA. At its peak sustained winds speeds reached 281 km/h with gusts exceeding 344 km/h. High winds and a central pressure of just 902 mb created a storm surge of 8–9 m, which breached the levées separating New Orleans from Lake Pontchartrain.
	Exposure was increased by the 9.5 million people living in coastal counties along the Gulf of Mexico between Louisiana and Florida, including large metropolitan areas such as New Orleans, Tampa and Baton Rouge. Rapid population growth has occurred in this area since 1950 (350%), increasing the number of people at risk.
	New Orleans is particularly exposed to storm surges. It is a city surrounded by water (Lake Pontchartrain and the Mississippi River) and large parts of the site are below sea level.
Vulnerability	Katrina's progress was monitored and tracked as it developed from a tropical depression to tropical storm and finally to a category 5 hurricane. Before Katrina made landfall, the National Weather Service issued hurricane watches and warnings. Accurate predictions of the location and time of landfall were made.
	President Bush declared a state of emergency in Alabama and Mississippi 2 days before Katrina made landfall and the governor of Louisiana declared a state of emergency on 26 August. On 28 August the mayor of New Orleans ordered an evacuation of the city. The city's superdome sports arena and conference centre were designated 'refuges of last resort' for the 150,000 people unable to flee the city. Although these measures did little to reduce the physical and economic damage caused by the storm, they probably saved thousands of lives.
	New Orleans' 560 km levée system protects the city from flooding but was built to withstand only a category 3 hurricane. Despite the risks, the levées had been poorly maintained. In 2004 the US Army Corps of Engineers had asked the federal government for $105 million to strengthen the New Orleans flood defences. It got just $40 million.

Impact	Flooding caused by the storm surge was made worse by the loss of wetlands in the Mississippi delta. Over the years large areas of wetland and salt marsh have become open water. In the past these wetlands acted as a buffer, absorbing water and giving protection against storm surges and flooding. The main cause of wetland losses is subsidence due to the extraction of natural gas in the delta. The levée system also played a part, preventing the Mississippi River from flooding and thus starving the delta of new sediment which would raise the level of the land surface.
	Following the Katrina disaster, a multi-billion dollar scheme got underway to strengthen New Orleans' flood defences. The aim is to raise the levées and flood walls around the city to a level capable of withstanding a category 5 hurricane by 2011. New water pumping stations are being built to remove water from the city. However, some experts question whether protecting areas of the city up to 2 m below sea level is sustainable in the long term. Rising sea level and the sinking delta may eventually mean that these areas have to be abandoned. By 2009, 31% of all houses in New Orleans were still uninhabited and the city's total population was only 60% of that before Katrina.

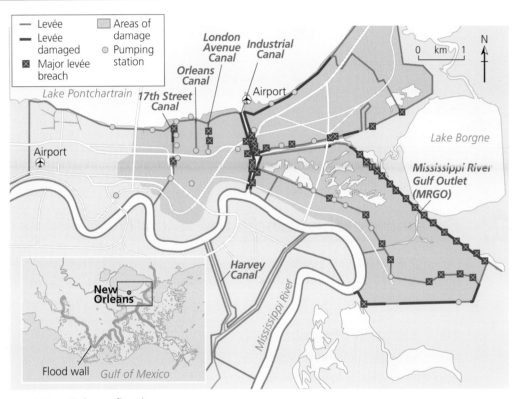

Figure 7.10 New Orleans flood map

9 Tornadoes

A **tornado** is a violently rotating column of air that reaches to the ground from a thunderstorm. Its most characteristic feature is a **funnel**, made visible by dust and water droplets, which extends towards the surface (Figure 7.11). Wind speeds within the funnel may reach 500 km/h. Tornadoes are localised and highly destructive. Destruction may be confined to a belt no more than 50 m on either side of the storm track. Beyond this belt, adjacent areas are often completely unscathed.

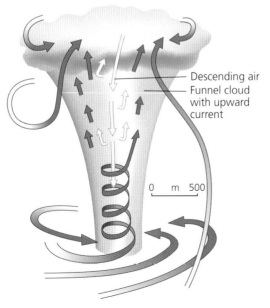

Figure 7.11 Structure of a tornado

No other region suffers so many destructive tornadoes as 'tornado alley' in the US Mid-West, between the Rockies and the Appalachians. Tornadoes are most frequent in spring and early summer, when cold dry air from the Rockies collides with warm humid air from the Gulf of Mexico. As these contrasting air masses meet, the warm humid air is forced upwards and forms huge cumulonimbus clouds and thunderstorms. Occasionally, thunderstorms group together to form **supercells**. The most dangerous tornadoes originate in these supercells, which also produce intense rainfall and damaging hail showers.

Box 3 *Development of tornadoes*

Two air masses — one cold and dry, the other warm and humid — collide. The boundary between the air masses forms a cold front. Sharp contrasts in pressure exist on either side of the front, which cause the wind to veer abruptly and change direction with height (**wind shear**).

Cold, dry air lies above the warm humid air. Powerful updraughts develop as warm air becomes unstable. In response to wind shear, rapidly rising air starts to spin and a **vortex** develops in the middle of the thunderstorm.

Eventually, in a process that is not fully understood, the vortex extends downwards from the cloud base to the surface as a tornado.

9.1 The impact of tornadoes

In an average year, about 1,000 tornadoes are reported across the USA. They cause around 80 deaths and over 1,500 injuries. Damage from tornadoes is caused by violent winds and flying debris. Wind speeds of up to 500 km/h can pick up cars and rip houses to shreds. In extreme storms, damage paths can exceed 1.5 km in width and 80 km in length.

10 *Extreme weather conditions* _ UNIT 2

Extreme weather conditions are exceptional events such as storms, droughts, heat waves, severe spells of cold weather and heavy rainfall and snowfall. These events occur infrequently and they are often hazardous.

10.1 Blizzards and cold spells

Heavy snowfalls in the British Isles occur during spells of unusually cold weather. One such cold snap occurred between 27 February and 7 March 2006.

CASE STUDY 3	Heavy snowfall in northern Britain, 27 February to 7 March 2006
Background	Low pressure over Scandinavia fed cold arctic maritime air southwards across Britain (Figure 7.12). Warmed through contact with the sea, this unstable air mass swept south. Several polar lows formed in the air stream, bringing widespread snowfalls to Scotland and England.

Figure 7.12 Synoptic chart, 1 March 2006

Snowfall	Heavy snowfalls were recorded in northeast Scotland on 2 and 3 March. At Glenlivet in Banffsh..., the snow depth was 29 cm. Significant accumulations also occurred in the coastal areas of eastern England. Inland, the snow showers died away as the air mass cooled on contact with the cold ground surface. Therefore, most of central and southern England was dry with clear skies, long sunny spells and night frost. However, on the snowfields of Scotland, clear night skies led to extremely low temperatures (−16.4°C at Altnaharra and −12.7°C at Braemar).
Impact	Scotland bore the brunt of the cold spell. Four successive days of blizzard caused severe disruption to transport and brought much of northern Scotland to a standstill. Roads were affected by deep drifts and reached record depths at Aberdeen. Hundreds of motorists were stranded and their vehicles abandoned on the A96 between Inverness and Aberdeen. Central Aberdeen and Inverness were gridlocked on 3 March. Domestic refuse collection was suspended in Aberdeen and shops and banks closed early. The snow also affected rail and air transport. Several trains became stuck in deep drifts on the Aberdeen to Dundee line. Inverness airport was closed and flights from Aberdeen airport were severely disrupted. Throughout northern Scotland, hundreds of schools closed. Most Scottish football fixtures were postponed.

11 *Anticyclones*

Anticyclones are areas of high pressure that affect the weather in mid-latitudes around one day in four. They sometimes bring extreme temperatures and drought. Once established, they can persist for days or even weeks, making conditions even more extreme.

Box 4 *Anticyclones in middle to high latitudes*

Figure 7.13 shows that in anticyclones air subsides throughout the lower atmosphere (**troposphere**). As the air sinks it warms, leaving the troposphere cloud-free. The result is dry weather. Clear skies bring lots of sunshine and night-time frost in winter. However, if the sinking air diverges before reaching the surface, a shallow **temperature inversion** will develop near the ground. Stratus clouds often fill the inversion layer and create overcast conditions. In winter **radiation fog** is common at night; trapped in the inversion layer it may persist all day. Anticyclones often bring exceptional weather conditions in middle to high latitude areas with extreme temperatures, low rainfall and drought.

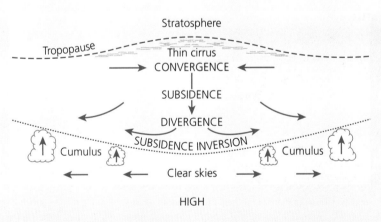

Figure 7.13 Structure of an anticyclone

11.1 Anticyclonic blocking

In western Europe, **anticyclone blocking** occurs when a large, slow moving anticyclone becomes established over the continent and disrupts the normal westerly flow. Blocking anticyclones force mild and humid Atlantic air further north or south of its usual track and airflow becomes more meridional. Northerly or southerly airstreams dominate the weather, often bringing extreme temperatures. Northerly

flows introduce polar and arctic air, with below average temperatures in all seasons. A southerly flow brings tropical air from North Africa and above average temperatures. In summer this may give **heat wave** conditions. Blocking situations are often responsible for droughts and can result in unusually high or low sunshine levels.

CASE STUDY 4 Heat wave in Europe, 2003	
Cause	An exceptional and prolonged heat wave struck Europe in July and August 2003. The heat wave was caused by an anticyclone anchored over northern France. For 20 days it blocked Atlantic air masses and drew in extremely hot dry air from North Africa.
Extreme weather	Extreme heat affected southern England and the Midlands. On 10 August the highest-ever maximum temperature (38.5°C) was recorded in the UK. At Gatwick in southeast England, daily maximum temperatures exceeded 25°C on ten successive days between 3 and 13 August. Average temperatures hit record levels on the continent. In France and Italy temperatures soared to 40°C and remained unusually high for 2 weeks. In southern Spain temperatures in excess of 40°C were recorded in most cities, peaking at 46.2°C in Cordoba.
Impact	Western Europe was unprepared for the 2003 heat wave — for example, few homes had air conditioning. Extreme heat, especially in large urban areas, created a major health crisis. Some 35,000 deaths, most of them elderly people, were attributed to the 2003 heat wave. Worst hit was France with 15,000 deaths. The heat also claimed 7,000 lives in Germany, 2,000 in the UK and 1,400 in the Netherlands. Mortality was highest in large urban areas where heat-absorbent surfaces and sparse vegetation cover amplified temperatures. In France the highest mortality rates were in the major cities. Around one-third of the excess deaths in France were in the Paris region and 80% of the victims were aged 75 and over.
	The combination of excessive heat and drought had severe effects on economic activities, especially agriculture. The heat wave was accompanied by drought, which seriously reduced crop yields. Wheat yields dropped by 20% in France, 13% Italy and 12% in the UK. Ukraine had a 75% reduction in its wheat harvest, and in Moldova the decline was 80%. Shipping was suspended on the Elbe and Danube due to low river levels. Melting glaciers caused avalanches and flash floods in Switzerland. Portugal had massive wildfires that destroyed 10% of the forest area and killed 18 people.

Box 5 *Killer heat waves and cities*

Heat waves mainly affect the elderly, the young, the chronically ill and urban dwellers.

Normal body temperature is 37°C. When ambient temperatures rise, the human body maintains its ideal temperature by perspiring and varying blood circulation. When the internal body temperature rises above 40°C, internal organs are at risk: if the body temperature is not reduced, death follows.

High humidity makes extreme heat even more dangerous. With little or no evaporation to cool the body, perspiration becomes ineffective.

Urban dwellers are most at risk during heat waves. Urban surfaces with low albedos (surface reflectivity) absorb the heat and create 'heat islands'. This is particularly evident at night when heat stored by the urban fabric during the day is released. Urban areas also produce heat through domestic heating, factories and vehicles. The lack of cooling vegetation in urban areas adds to the high temperatures. On some days in summer the difference in temperature between a large city such as London and the surrounding countryside can be as much as 10°C. A lack of moisture in cities also reduces evaporation and cooling and gives more energy to heat the atmosphere. Apart from heat, people's bodies may be stressed by a pollution 'dome' that settles over cities during anticyclonic weather.

12 *Depressions*

Depressions are large, travelling low pressure storms that dominate the weather in northwest Europe. Part of the atmosphere's general circulation, they transfer warm air from the tropics to the polar regions. Depressions bring windy, wet and changeable conditions. Within depressions, warm and cold air meet

along boundaries known as **fronts**. At these frontal zones warm air is forced aloft to form organised bands of thick cloud and heavy precipitation (Figure 7.14).

Ci = cirrus   Ns = nimbo stratus
Cs = cirro stratus   Sc = strato cumulus
Ac = alto cumulus   Cu = cumulus
As = alto stratus   Cb = cumulonimbus

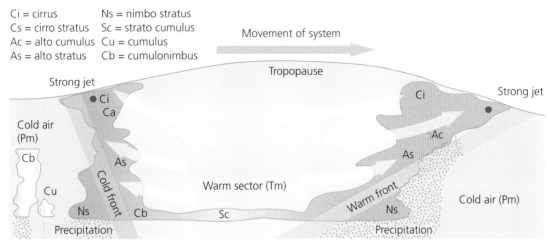

Figure 7.14 Formation of a depression

CASE STUDY 5   Record rainfall in the UK, 2007

Cause	In June and July 2007, the UK's weather was dominated by a succession of Atlantic depressions. The cause was the persistent location of the jet stream across central Britain. The jet stream, a fast moving ribbon of air 7–8 km above the ground, controls the formation and movement of depressions. Normally high pressure from the Azores spreads northwards across the UK in the summer, giving settled weather conditions. This failed to happen in 2007.
Rainfall	The summer of 2007 in the UK was the wettest on record. With the exception of northern Scotland, all regions had above average rainfall. In June, rainfall totals in some regions (notably Yorkshire) were three times the monthly average and in Worcestershire in July rainfall was four times the monthly average.
Impact	Exceptionally heavy rains caused flooding, damaged crops and disrupted farming in many parts of the UK. Harvesting crops like vining peas was impossible in Yorkshire and the East Riding, where 60% of the crop was destroyed. Waterlogged soils caused potato blight and much of the potato crop simply rotted in the ground. Around 42,000 ha of farmland were flooded. Flooded grasslands reduced the silage crop and wheat was badly affected. Lost farm production was between £11 million and £24 million. In the winter of 2007/8 shortages of potatoes, peas and cereals led to price increases in the supermarkets. Poor summer weather affected tourism. International visitors to the UK fell from 3.7 million in August 2006 to 3.2 million in August 2007. Part of this decline was blamed on the dismal weather. However, some regions were particularly badly hit. In Gloucestershire and Worcestershire, flooding closed large tracts of the countryside, seriously reducing visitor numbers and damaging the local economy. Tewkesbury Abbey had barely one-quarter of its expected visitor numbers and both the cricket ground and racecourse at Worcester were closed for several weeks. Many British families booked last-minute holidays to southern Europe to escape the rains.

1 Ecosystems

Ecosystems are communities of plants, animals and other organisms (the **biotic** component) and the environment (the **abiotic** component) in which they live and interact (Table 8.1). The physical environment provides the energy, nutrients and living space that plants and animals need to survive.

Table 8.1 Components of ecosystems

Abiotic components	Biotic components
Rocks	Plants (primary producers)
Relief	Animals (herbivores, carnivores, omnivores)
Atmosphere (gases) and climate	Detritivores (fungi, microbes etc.)
Soil	People
Water	
Solar radiation	
Fire	
Gravity	

Ecosystems exist at different scales, from a single oak tree to the Amazon rainforest and even the Earth itself. Those ecosystems that extend over large geographical areas (e.g. the tundra) are known as **biomes**. Most ecosystems are **open systems** (Figure 8.1). This means that both energy (e.g. solar radiation) and materials (e.g. minerals from weathered rocks, and water) cross ecosystem boundaries.

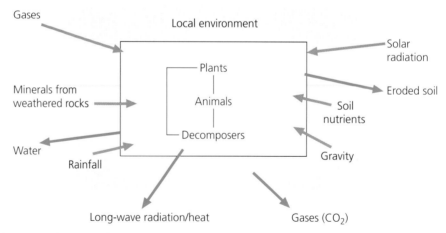

Figure 8.1 Inputs and outputs in an open ecosystem

Plants and animals interact with the physical environment and with each other. These interrelationships bind the components of ecosystems into a coherent whole. We refer to this quality of wholeness as **holisticity** (Box 1).

Box 1 Interrelatedness in ecosystems

Because ecosystems are holistic, change in any one component often has far-reaching consequences. Therefore, a hill sheep farmer who increases the density of sheep on upland pastures may inadvertently cause damage to the ecosystem. Where densities exceed the **carrying capacity**, overgrazing may destroy the vegetation cover, increasing runoff and causing accelerated soil erosion. The resulting increases in suspended sediment in local streams may degrade freshwater habitats, destroying insect larvae and other animals that depend on them further up the **food chain**.

Sunlight, captured by the leaves of green plants, is the primary energy source for most ecosystems. This energy is then transferred along food chains and interlocking food chains known as **food webs**. The flow of energy within ecosystems occurs in a number of stages or **trophic levels** (T).

● T1: green plants intercept sunlight and, in the process of photosynthesis, convert sunlight, water, carbon dioxide and mineral nutrients into carbohydrates. Green plants are the primary producers (**autotrophs**) in ecosystems.

● T2: plant-eating animals or herbivores convert some of the energy into animal tissue. Herbivores are the primary consumers in food chains. Consumers are also known as **heterotrophs**.

● T3: meat-eating animals or carnivores prey on herbivores. Carnivores, occupying the third tropic level in a food chain, are secondary consumers.

● T*n*: at the end of each food chain, there is a top or apex predator. Depending on the length of the chain, this animal may be a tertiary or quaternary consumer.

● At each trophic level, detritivores such as fungi and microbes decompose dead organic matter and animal faeces, consuming energy and releasing gases (carbon dioxide, methane) and mineral nutrients.

At each trophic level in a food chain there is a loss of energy. This is because organisms convert only a small proportion of the energy they consume into living tissue. Most is used to keep the organism alive and is lost as heat in **respiration**. As a result:

● there are limits to the number of trophic levels in food chains
● there is a reduction of biomass at each trophic level, as shown in Figure 8.2
● there is a reduction in the population of animals at each trophic level

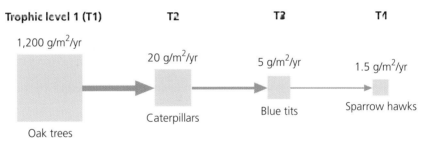

Figure 8.2 A simple food chain in an oak woodland

Ecosystems in harsh environments, such as deserts (Figure 8.3) and tundra, are vulnerable to change by human activities. These ecosystems are fragile because of their relatively simple structures. They have few species, low rates of plant growth, slow cycling of nutrients and short food chains. In contrast, complex ecosystems such as the tropical rainforest are far more robust and able to accommodate change.

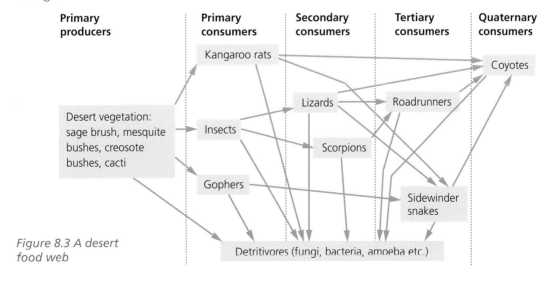

Figure 8.3 A desert food web

Most natural ecosystems can adjust to change and temporary disequilibrium through the operation of negative feedback loops. However, such loops that restore stability in ecosystems are easily destroyed through human activities.

Box 2 *Dynamic equilibrium*

Most natural ecosystems, unaffected by people, are in a state of dynamic equilibrium. They are dynamic (rather than static) in the sense that they have continuous inputs, throughputs and outputs of energy and materials. From year to year there may be small fluctuations in animal populations and plant productivity. However, in the long term they remain stable, showing no fundamental changes over time.

Box 3 *Change and feedback in ecosystems*

Change in ecosystems produces two responses: positive feedback and negative feedback.
■ **Positive feedback** occurs when change leads to further change — a kind of snowball effect. In ecological succession, some plant species change the soil and microclimate. This allows new species to invade. They in turn modify the environment until other species take over and become dominant.
■ **Negative feedback** leads to self-regulation, which restores ecosystems to balance. In the Arctic tundra, a glut of berries and edible seeds in a particular year may cause large increases in the population of small rodents. Animals such as arctic foxes and snowy owls respond to this abundance of prey by increasing their numbers. By predation, these carnivores restore balance between the rodent population and the environment.

2 *Nutrient cycles*

Nutrients are the chemical elements and compounds needed by plants and animals. They cycle between the living organisms and the physical environment within ecosystems. There are three sources of nutrients:
● Rocks are the source of most nutrients. On weathering, rocks release nutrients such as potassium, calcium and sodium into the soil, where they are absorbed by the roots of plants.
● Plants obtain some nutrients, such as nitrogen and carbon, directly from the atmosphere. Some mineral nutrients are also dissolved in rainwater. Some tropical rainforest plants (epiphytes) with aerial roots rely solely on the mineral nutrients in precipitation.
● Eventually, most mineral nutrients return to the soil as plant litter. The dead leaves, roots and stems are decomposed and mineralised by fungi and microbes in the soil.

Box 4 *Primary productivity in ecosystems*

There are two measures of primary productivity in ecosystems:
■ **Gross primary productivity** (GPP) is the amount of energy fixed in photosynthesis. It is measured in grams/m²/year.
■ **Net primary productivity** (NPP) is the amount of energy fixed in photosynthesis minus the energy lost in respiration. NPP is also measured in grams/m²/year. The net primary productivity of selected biomes is given in Table 8.2.

Table 8.2 Net primary productivity of selected biomes

Biome	NPP (grams/m²/year)
Tropical rainforest	2,200
Temperate deciduous forest	1,200
Savanna grasslands	900
Boreal coniferous forest	800
Temperate grassland	600
Tundra and alpine	140
Desert and semi-desert	90

Shortages of nutrients in an ecosystem may limit plant growth and influence species composition. For example, soils that contain little free calcium often support a more impoverished flora than neutral or slightly alkaline soils.

Human activity may interrupt nutrient cycles and deplete nutrients in ecosystems. Deforestation often results in the **leaching** of mineral nutrients from the soil.

Nutrient cycles vary in their speed. In the tropical rainforest, high temperatures and humid conditions speed up decomposition and cause rapid nutrient cycling (Figure 8.4).

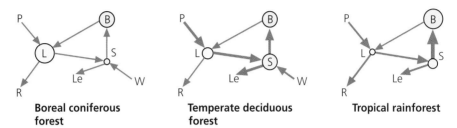

Boreal coniferous forest   **Temperate deciduous forest**   **Tropical rainforest**

B = Biomass   L = Litter   S = Soil   P = Precipitation   R = Runoff   Le = Leaching   W = Weathering

○ Size is proportional to the amount of nutrient stored

→ Width equals nutrient flow as a percentage of the nutrients stored in the source

Figure 8.4 Nutrient cycles in three forest ecosystems

In the boreal coniferous forest, low temperatures slow down the activities of the decomposers. As a result, several years of leaf litter accumulate on the forest floor. Sometimes the nutrients tied up in the litter rely on wildfires to release them.

The richness of a nutrient cycle depends partly on the vegetation. Most conifers in the boreal forest, such as spruce, pine and cedar, require few nutrients. Unused nutrients weathered from rocks are quickly leached from the soil. Oak trees, by comparison, are more demanding. By absorbing many more nutrients from the soil they sustain a much richer nutrient cycle.

3   Ecological succession

The sequence of vegetation changes on a site through time is called **ecological succession**.
- Primary succession describes vegetation changes on sites previously uncolonised (e.g. sand dune, mudflat, bare rock). Primary succession in a deglaciated area is shown in Figure 8.5.
- Secondary succession describes vegetation changes on sites where the original vegetation cover has been destroyed (e.g. the fires in Yellowstone National Park, USA in 1988, which destroyed extensive areas of lodgepole pine).

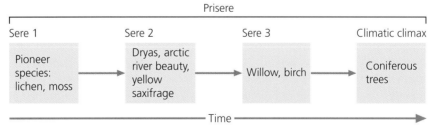

Figure 8.5 Primary succession in a recently deglaciated area

Many ecological successions have their greatest biodiversity and productivity at an early stage. In forest successions, an early stage of diverse perennial herbs, grasses and ferns is eventually shaded-out by forest trees. However, productivity is often highest when young saplings are growing vigorously. At

the climax stage, most **biomass** comprises non-productive supporting structures such as trunks and branches.

Ecological succession has the following characteristics:
- over time the physical environment, modified by plant growth, becomes increasingly attractive to a wider range of plant species
- there is a progressive increase in nutrient and energy flows
- biodiversity increases
- net primary production increases

Examples of primary succession in coastal dune and salt marsh environments are described in Topic 5.

4 The effect of human activity on natural ecosystems

Human activities often modify (and degrade) natural ecosystems by:
- reducing biodiversity (number of species per unit area)
- eradicating those plants and animals that compete with people for food (e.g. weeds and pests)
- extinction (most often through habitat destruction) of animals with small populations (e.g. at the end of food chains)
- reducing flows of energy and the amount of sunlight photosynthesised
- impoverishing nutrient cycles, with reductions in inputs of plant litter and increased losses of nutrients to leaching
- accelerating soil erosion by wind and runoff
- maintaining ecosystems at an early successional stage (e.g. grasslands rather than forest in upland Britain)
- introducing, either deliberately or inadvertently, alien species (e.g. cane toads in northern Australia, mink in the UK, Japanese knotweed and Himalayan balsam in the UK)
- fragmenting natural ecosystems and habitats
- polluting the atmosphere (e.g. acid rain, carbon dioxide and global warming), rivers (e.g. industrial effluent, domestic waste) and oceans (e.g. increased acidity)
- destroying habitats such as hedgerows, wetlands and ancient woodlands through intensive farming and urbanisation

TOPIC 9 — Population and resources

1 Population change

At the global scale, the number of births and deaths influence population change. If births outnumber deaths, the global population expands. This is **natural increase**. When deaths exceed births, the population declines and there is **natural decrease**. At continental, national, regional and local scales a third factor, **migration**, affects population change. The following formula summarises population change at these scales: Population change = (births − deaths) +/− migration

Box 1 Calculating rates of natural increase/decrease

% natural population change per year = (crude birth rate per 1,000 − crude death rate per 1,000)/10

For example:

% global natural increase in 2009 = (20 − 8.2)/10 = 1.18%

The time taken to double a population can be estimated from the natural increase rates thus:

Doubling time = 693/(natural increase rate × 10)

Therefore, at 2009 rates of natural increase it will take 693/11.8 = 58.7 years for the world's population to double.

2 Fertility and mortality

Fertility is the occurrence of live births. Many factors influence fertility including economic status, religion, government policies, female literacy, the economic value of children and the availability of contraceptive devices.

The **crude birth rate** (CBR) is widely used as a measure of fertility. The CBR is the ratio of the number of live births to the total population. It is usually expressed per 1,000 of the population. As a measure of fertility, the CBR has drawbacks because it is strongly influenced by age–sex structure. Regardless of how many children each woman produces, mature populations and populations with relatively few women will have low CBRs. However, the CBR is useful in calculating population change. Box 2 describes alternative measures of fertility.

Box 2 Other measures of fertility

General fertility rate: the number of live births in a year as a ratio of the number of women aged 15–44, expressed per 1,000 women. In 2007 the general fertility rate in England and Wales was 63.5 per 1,000.

Age-specific fertility rate: the number of live births to women in 5-year age groups per 1,000 women. Age-specific fertility for 25–29-year-old women in England and Wales in 2004 was 98.6 per 1,000, compared to 10.4 per 1,000 for women aged 40 years and over.

Total fertility rate (TFR): the average number of live births to women who have completed their families. The reproductive TFR (i.e. the rate needed just to replace the population) is 2.1. The global average TFR in 2009 was 2.58. This varied from 1.21 in Japan to 7.29 in Mali, West Africa.

Mortality is the occurrence of death. Rates of mortality depend on many factors such as age, diet, healthcare, economic status and disease. The **crude death rate** (CDR) is calculated in the same way as the CBR and has similar disadvantages. Age-specific mortality is similar to age-specific fertility. In developed countries it generally peaks in infancy and then declines until the mid-teens. Thereafter, there is a slow increase until extreme old age, when the rate rises very steeply. The **infant mortality rate** (IMR) is the number of deaths of infants under 1 year old per 1,000 live births. The IMR is a particularly sensitive indicator of the standard of living in a society.

2.1 Global population change

In 1999 the world's total population reached 6 billion. This figure reflects the explosive population growth of the past 200 years and especially since 1950 (Table 9.1). Population growth has resulted from an imbalance between births and deaths. While death rates in developing countries have fallen rapidly, birth rates have remained high. Natural increase peaked (at around 2.2%) in 1962/63. Since then, falling birth rates have reduced global rates of natural increase to around 1.1%. Even so, the global population expanded by nearly 90 million in 2009. Latest estimates forecast a total population of around 9.5 billion in 2050, with zero growth achieved only towards the end of the twenty-first century.

Table 9.1 World population growth, 1800–2009

Year	Population (bn)
1800	1.125
1850	1.402
1900	1.762
1950	2.556
2000	6.073
2003	6.263
2009	6.782

2.2 Spatial patterns of global population change

There are great spatial differences in the global pattern of population change. In the early twenty-first century, the main features of this pattern and the implications for future population growth are as follows:

- Very low growth rates throughout most of the economically developed world. Developed countries' share of world population will fall from 20% to 14% between 1995 and 2050. Europe's population will shrink by 90 million during this period (Figure 9.1). TFRs in much of Europe are already below replacement level.
- Rapid but declining rates of population growth in most developing countries. However, the youthful populations of most developing countries mean than, even with declining fertility, there is great momentum for future growth.
- Most world population growth between 1995 and 2050 will be concentrated in Asia, especially India, China and Pakistan. India (with a TFR of 2.72 in 2009) will overtake China (with a TFR of 1.79 in 2009) as the world's most populous state by around 2025.
- Very rapid population growth in most Islamic countries in south and southwest Asia (e.g. Pakistan, Iran) and in sub-Saharan Africa (e.g. Nigeria, Kenya).
- Moderate population growth in Latin America, where fertility levels are generally low and national populations are relatively small.

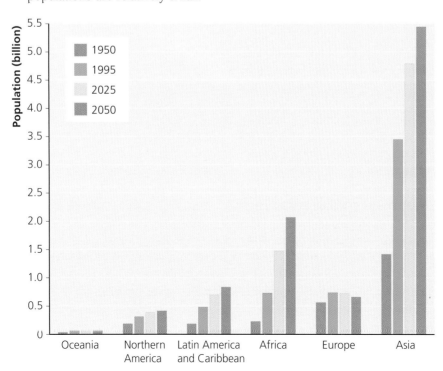

Figure 9.1 Changing distribution of global population by continent, 1950–2050

2.3 The demographic transition model

The **demographic transition model** describes the shift from high fertility and mortality to low fertility and mortality that occurred in Europe between 1750 and 1950. As Figure 9.2 shows, the transition involved four stages and was accompanied by economic growth and rising living standards. The model suggests that population change and economic development are causally linked.

- Stage 1: Pre-industrial. Fertility and mortality are high. There is little population growth. High fertility is needed to ensure the survival of the population. Children also provide a source of labour and security for parents in old age. Artificial contraception is unavailable. There is high mortality owing to poor nutrition, poor hygiene and lack of medical knowledge to combat disease.
- Stages 2 and 3: Industrial. Improvements in medicine and in economic, social and environmental conditions cause mortality rates to fall. In stage 2, fertility remains high and results in rapid population growth. The fertility decline evident in stage 3 lags behind mortality decline. Even so, the rate of population increase begins to slow. Fertility decline occurs because children become expensive (long education period), the state provides for security in old age, artificial contraception is available and infant mortality is low so that few children die in infancy.
- Stage 4: Post-industrial. Both fertility and mortality are at low levels. Population growth is either very slow or has ceased altogether. Further causes of low fertility include women postponing marriage until their mid-twenties; more educated women with high-status jobs; more unmarried couples living together who are less likely to have children than married couples; and more effective methods of birth control.

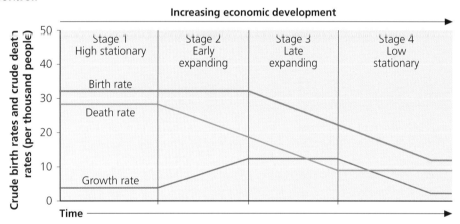

Figure 9.2 The demographic transition model

It is important to recognise that the demographic transition model is based on population changes that occurred in western Europe between 1800 and 1950. These changes were linked to economic growth, rising living standards, improvements in medicine and changes in societal values caused by urbanisation. However, the model does not describe accurately the experiences of most developing countries in the past 50 or 60 years. There are several reasons for this.

- Absolute numbers of population increase and growth rates have been far higher in developing countries than in nineteenth-century Europe. No countries in Europe had annual population growth rates of more than 1%.
- Mortality decline was a gradual process in nineteenth-century Europe. It followed improvements in living conditions brought about by economic growth and advances in tackling diseases such as smallpox. In developing countries, mortality decline has been more rapid. This decline is due largely to the application of modern medical techniques rather than to progress in living standards. In some countries, such as China and Bangladesh, government policies have speeded up population change. In southern Africa, the HIV/AIDS epidemic has caused an exceptional increase in mortality.
- Most developing countries are less urbanised than countries in nineteenth-century Europe. The economic advantages of large families (most people still live in rural societies), low levels of literacy (particularly among women) and (in the absence of rapid urbanisation) the survival of traditional cultures and values have proved obstacles to lowering fertility levels.

3 *Age–sex structure*

Age–sex structure is the composition of a population according to age groups and gender. We normally represent age–sex structure with a special kind of bar chart known as a **population pyramid**. Some examples are shown in Figure 9.3. Several factors influence the shape of population pyramids. Some are short term such as wars, epidemics and famines; others, such as fertility and mortality control, have a long-term effect. These long-term influences give rise to distinctive pyramid shapes.

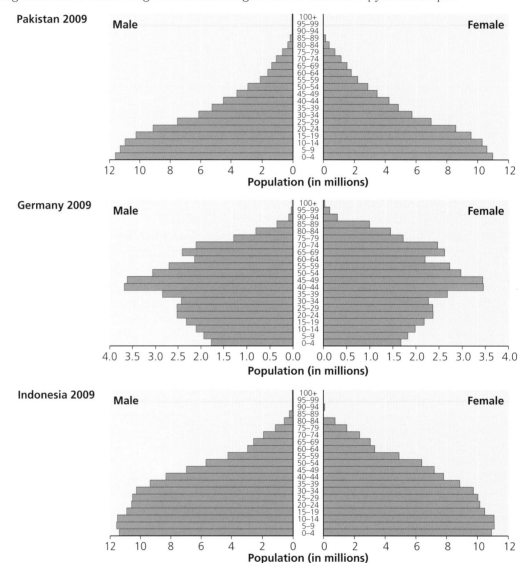

Figure 9.3 Population pyramids: (a) Pakistan, (b) Germany and (c) Indonesia, 2009

- Broad-based pyramids, such as Figures 9.3a and c, indicate youthful populations with large proportions of children and high levels of fertility. If fertility increases over time, the pyramid is progressive. Some pyramids show the effects of high fertility in the past. An unusually large birth cohort may follow a 'baby boom', such as occurred in western Europe at the end of the Second World War (Figure 9.3b). Today, these countries have unusually large numbers of older adults aged between 55 and 65 years.
- Rapidly tapering pyramids, again like Figure 9.3a and c, suggest high levels of mortality, with significant reductions in numbers at each 5-year age group. These populations usually have only small proportions of old people — hence their narrow apex.
- Straight-sided pyramids, with little reduction in the size of age groups between 0 and 60 years, suggest both low fertility and low mortality.

- Pyramids with a narrow base that broadens with age, such as Figure 9.3b, indicate recent reductions in fertility. These pyramids are known as regressive.

Age structure can also be measured by a number of indices, as shown in Table 9.2.

Table 9.2 Indices of age structure

Old age index	Dependency ratio	Juvenility index
Aged/adults	(Children and aged)/adults	Children/(adults + aged)

Migration has most effect on population structure at a local and regional scale. This is because migration is both age and sex selective. In most societies, the majority of migrants are young adults. Therefore, places of recent net migrational gain often have more young adults than average; those suffering a net migrational loss have fewer. The migrational effects of gender are more variable. In southern Africa, males migrate more often than females, giving unbalanced sex ratios both in receiving (usually urban) and sending (usually rural) areas. In South America, the situation is reversed, with females more often migrating than males.

3.1 The demographic impact of age–sex structure

The age–sex structure of a population has important implications for its future growth. Most developing countries have youthful populations — the result of high fertility and low mortality in the past 50 years or so. With so many children entering young adulthood in future, rapid population growth is almost inevitable. In this situation, we say that the population has considerable **momentum**. In China, total fertility fell to 1.8 children per woman in 2009. This figure is well below the replacement level of 2.1 children per woman. Yet population momentum in China is such that, even if total fertility remains at 1.8, the population will continue to increase until around 2035.

At a regional scale, migration often unbalances age–sex structure. Young adults are most likely to migrate from rural communities. This results in fewer babies and a general **ageing** of the population in rural areas. Ageing also occurs in communities that attract retirees. In the UK, the south coast is a popular retirement area, leading to ageing populations in towns such as Eastbourne and Christchurch. The destinations that attract young adult migrants (e.g. large cities in developing countries) experience the opposite effect: in-migration inflates the reproductive age groups and increases rates of population growth.

3.2 The economic impact of age–sex structure

Age–sex structure has important economic effects on the size of the workforce and **dependency**.
- The economically active population in developed countries are adults between the ages of 18 and 65 years. Already, low fertility in developed countries in the past 50 or 60 years has caused labour shortages and raised concerns about tax revenues. All of this has implications for the economy and for the provision of healthcare services and state pensions for older people. Partly in response to this situation, the UK government between 1997 and 2008 encouraged immigration, especially from EU countries in eastern Europe. However, the problem of 'greying' populations and increasing dependency is widespread in developed countries. In Japan, where total fertility rates are just 1.21 and the population actually fell in 2008, the absence of any significant immigration has made the issue of dependency especially urgent (Figure 9.4).
- An increase in the size of the dependent population also puts economic pressure on governments. In developing countries, high fertility has led to ever-increasing numbers of children and soaring demand for healthcare and educational services. In developed countries, the problem is the increasing proportion of old people and how to provide state pensions and extra resources for healthcare. In Germany, for example, the proportion of those aged over 65 years old increased from 15.8 to 20.5% between 1998 and 2010. Most developed countries have still to solve the problem of their 'greying' populations.

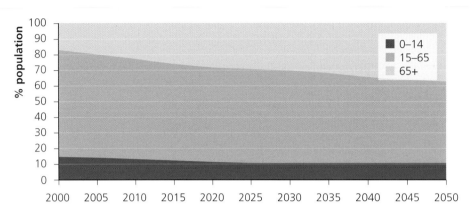

Figure 9.4 Japan's ageing population, 2000–2050

Legend: ■ 0–14 ■ 15–65 □ 65+

Migration

Migration is the permanent or semi-permanent change of residence of an individual or group of people. **Net migration** is the difference in numbers of in-migrants and out-migrants.

● When in-migrants exceed out-migrants, there is a **net migrational gain**.
● A **net migrational loss** occurs when there is an excess of out-migrants over in-migrants.

Migration, together with fertility and mortality, determines the population growth and population structure of a country or area.

4.1 Types of migration

Table 9.3 lists types of migration. Rural–urban and urban–rural migration have been responsible for significant shifts in population distribution in many countries. In developing countries, the net migrational gain of urban areas at the expense of rural areas results in **urbanisation**. Meanwhile, movements in the opposite direction in developed countries have led to **counter-urbanisation**. Some migrations, such as **stepwise migration** and **chain migration**, do not fit the simple classification of Table 9.3. Stepwise migration describes movement through a settlement hierarchy (e.g. from villages to small towns and cities). Chain migration links migration flows to kinship ties between, for example, a city and a particular rural region.

Table 9.3 Types of migration

Criterion	Description
Scale	International, inter-regional, intra-urban
Direction	Rural–urban, rural–rural, urban–rural, urban–urban
Distance	Long distance, short distance, regional/international
Decision making	Forced, voluntary
Causes	Economic, social, political, environmental

4.2 Causes of migration

Many factors influence the decision to migrate. At the simplest level, we divide these factors into two groups: push and pull.

● **Push factors** are the negative aspects of the current place of residence. They may be economic, social, political and environmental. They include lack of employment, low wages, poor housing, poor educational opportunities, political persecution and war.
● **Pull factors** are the attractions of places of destination. Often they are the inverse of push factors: better employment and educational opportunities, better housing, higher wages etc.

If the perceived push or pull factors are strong enough to overcome forces of inertia (cost of moving, disruption of social networks etc.), migration is likely to occur.

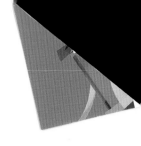

4.3 Perception and migration

At the level of the individual, **perception** has a strong influence on migration. Perception is a person's subjective view of the environment, derived from personal experience, the experience of others, the media etc. Through perception a potential migrant builds up a **mental image** of a destination. This mental image is often distorted and partial. Even so, it is this image, rather than objective reality, that is the basis of decision making. Migrants' mental images often fail to accord with reality. This is one reason why for every migration there is a movement in the opposite direction — a counter-movement of disillusioned migrants back to their place of origin.

4.4 Lee's migration model

Lee's model, shown in Figure 9.5, develops the ideas of push and pull factors and perception. According to Lee, potential migrants assess the **place utility** of their current place of residence and of their potential place of destination. Place utility depends on the migrant's perceptions and mental images. Both origin and destination have positive and negative attributes. These positive factors in the place of destination are the same as pull factors. Negative attributes in the place of origin are push factors.

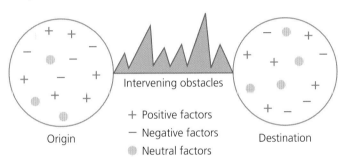

Figure 9.5 Lee's model of migration

Origin Destination

Intervening obstacles

+ Positive factors
− Negative factors
⬤ Neutral factors

For example, a redundant farm worker might regard the lack of employment opportunities in his village as a negative attribute, but the social network of the rural community as a positive one. Other factors might be neutral and not influence the migration decision. If the farm worker assesses the place utility of a possible destination to be greater than the present location, the decision is made to migrate.

Lee's model also takes account of another factor: intervening obstacles. These may include physical distance or physical obstacles (poor communications, cost of movement, mountain barriers etc.) between place of origin and place of destination; political obstacles, such as international borders; legal obstacles, such as problems of obtaining work permits etc.

4.5 Consequences of migration

Migration affects populations unequally: some people and groups are more likely to migrate than others. In other words, migration is a selective process influenced by factors such as age, gender, education, occupation and stage in the family cycle. The selectivity of migration may cause unbalanced age structures and sex ratios in both receiving and sending areas.

Because young adults are more likely to migrate than older adults and the aged, most populations suffering a net migration loss experience ageing. Populations undergoing net migration gain often experience the opposite effect, with disproportionate numbers of young adults and children. However, migration on retirement, which is increasingly common in developed countries, creates a top-heavy population pyramid with a large proportion of old people.

The effect of migration on sex ratios varies with culture. In South America women are more likely to migrate than men, whereas in sub-Saharan Africa men form the bulk of migrants. Other features of migration include the greater mobility of more educated people and the influence of critical stages in the family cycle on the decision to migrate (e.g. young adults becoming independent, leaving home or starting a family).

Regions experiencing heavy net migrational loss may suffer an absolute decrease in population. This phenomenon is known as **depopulation**. It has affected many remote rural areas in western Europe in the last 150 years. In areas such as northwest Scotland and the Massif Central in France, it even continues today.

CASE STUDY 1	Immigration to the UK: 2004–07
Background	There was an annual net migration gain in the UK for most of the period 1990–2007. Before 2004, most immigration originated from Africa, the Middle East and south Asia. Since the enlargement of the EU in 2004, the majority of immigrants have come from EU countries in eastern Europe (e.g. Poland, Lithuania). Between 2004 and 2006, 181,000 immigrants came to the UK from eastern Europe and in 2007 the overall net migration gain from international migration was 237,000. Immigration has been at record levels for the past decade and is currently the main driver of population growth (Figure 9.6).
Causes	The bulk of legal immigration is driven by economic factors. Citizens of other EU countries have a legal right to live and work in the UK. Given the huge economic disparities that exist between the UK and eastern Europe and the relaxed approach to immigration by the British government, large-scale immigration is not surprising. Immigration is driven by a combination of economic push and pull factors. In 2007, GNI per capita in the UK was US$40,660, compared to US$9,850 in Poland; unemployment in Poland in March 2007 was twice as high as in the UK; and the minimum wage in the UK was more than five times greater.
Geographical distribution	Unlike immigration from Africa, the Middle East and south Asia, which was geographically concentrated in a few regions (London, the southeast, Yorkshire and the Humber), east European immigrants are widely dispersed in large cities, market towns and even rural areas.
Impact	Immigrants provided a source of skilled and relatively cheap labour which the British economy needed in the economic boom years of 2004–07. However, a report by the Select Committee for Economic Affairs in 2008 suggested that the overall benefits to the UK economy had been exaggerated and that the benefits depend on the skills of the immigrants. Many immigrants are young adults with families and this has put a strain on housing and public services such as schools and healthcare in some parts of the UK. The youthful profile of immigrants will increase the birth rate, helping to raise the UK population to 71 million by 2030. In 2008 nearly 1 in 4 births in the UK was to women born overseas. There are environmental concerns over rising population pressure, with ever-increasing demands for housing and the loss of countryside to urban development.

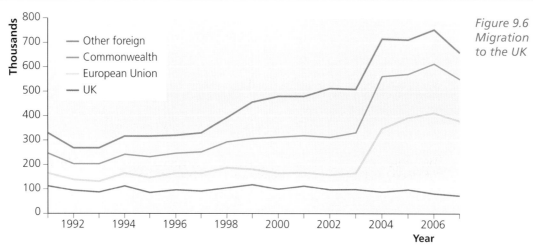

Figure 9.6 Migration to the UK

5 *Population policies*

Most population policies in developing countries aim to tackle the problem of rapid population growth (Case Study 2). Policies that aim to do this by reducing fertility are known as **anti-natalist**. For economic and political reasons, a few countries (e.g. Malaysia) have **pro-natalist** policies designed to increase population.

6

CASE STUDY 2   China's population policy

China's controversial anti-natalist population policy was a response to perceived overpopulation in the 1970s. Between 1950 and 1980 China's population increased from 560 to 985 million. Such growth threatened shortages of food, fresh water, fossil fuels and other natural resources. At the time, China's optimum population was estimated at 700 million. As a communist country, China gives priority to state interests over the rights and freedom of individuals. Therefore, the government was able to impose an authoritarian population policy that would be impossible in democratic society.

China introduced its one-child policy in 1979. Women who had more than one child incurred economic penalties or were coerced into abortions. The legal age for marriage was increased to 22 years for men and 20 years for women. The policy had most success in towns and cities, where it was easier to enforce and where small families were more acceptable. However, in rural areas the policy met considerable resistance: it was difficult to enforce and explain to poorly educated farmers. Therefore, in rural areas it was applied less rigidly. For example, if a couple's first child was a girl, they might be allowed try for a boy (in Chinese society there is a preference for male offspring — inheritance is through the male line; on marriage, wives live with their husband's family and provide extra labour, a dowry and support for her parents-in-law in old age). Even in urban areas there has now been some policy relaxation. If a couple were both only children, they could try for a second child 4 years after the birth of their first child.

The impact of China's population policy has been dramatic. Whereas the country's population increased by 73% between 1950 and 1979, growth fell to 37% between 1979 and 2008. Today, one-third of all Chinese families are single-child families. It is claimed that the policy has been responsible for 400 million fewer births.

However, the effect of the policy on China's age structure will give rise to economic and social problems in future.
- The proportion of young people has fallen steeply, threatening labour shortages in cities.
- Thanks to improvements in healthcare and living standards, the proportion of old people has risen, increasing levels of dependency. Although only 9% of China's population was over 60 years old in the early 1990s, by 2030 the proportion will reach 25%. With little state provision for pensions and retirement benefits, the burden of looking after old people will fall on today's single child.
- The preference for male children has led to female infanticide and selective abortions of girls. The resulting gender imbalance will eventually mean a shortage of marriageable women, threatening the tradition of universal marriage.

6   *Natural resources*

Natural resources are 'stocks of physical assets that are not produced goods, and that are valuable to humans' (African Development Report, 2007). There are several classifications of natural resources, but the most widely used in geography is based on their renewability (Figure 9.7).

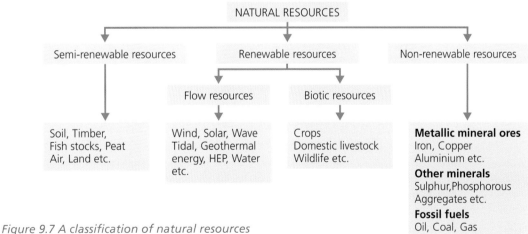

Figure 9.7 A classification of natural resources

6.1 Renewable resources

Renewable resources regenerate on human timescales through ecological cycles. Annual and fast-growing plants and crops are renewables. So too are domestic livestock and wild animals. Resources that also regenerate but over longer, intermediate timescales are known as **semi-renewables**. They include forests, fish stocks and soils.

A feature of renewables is their linkage with other resources within ecosystems. For example, wildlife and forests rely on water and soil for their renewal. Currently, the key challenge concerning renewables is their **sustainable management**. To achieve this, consumption must not exceed rates of regeneration.

Flow resources are a separate group of renewables. They have a permanent source and do not need regeneration. All forms of alternative energy, such as wind, solar and hydroelectric power, are flow resources. Although flow resources cannot be depleted, their exploitation depends on other natural resources such as energy, materials and space for construction.

6.2 Non-renewable resources

Non-renewable resources are finite on human timescales. Renewability of fossil fuels, metal ores and minerals such as gypsum and limestone is either zero or only over tens of millions of years. Thus the world's reserves of non-renewables are being steadily depleted, although in the case of metals such as steel and copper recycling can prolong their availability.

6.3 Technology and the changing definition of resources

Resources are defined by technology and societal values. Materials only become resources when the technology exists to find, exploit, process and use them. Palaeolithic hunter-gathers in Europe had a simple technology that relied on resources such as wild animals, plants, flint and wood. Without the technology to extract and smelt metal ores, it would be several thousand years before iron, copper and tin were recognised as resources.

7 Population and resources

The concepts of **optimum population**, **overpopulation** and **underpopulation** are relative ones that refer to the relationship between the size of a population and available physical resources. The number of people that can be supported sustainably by available resources is the **carrying capacity**.

- Optimum population: the population that maximises the population to resource ratios at a sustainable level (Figure 9.8).
- Overpopulation: an excess of people that reduces resources per person, resulting in lower living standards and which, in rural societies, often results in environmental degradation.
- Underpopulation: too few people to develop fully the resources of an area and raise standards of living to the optimum level.

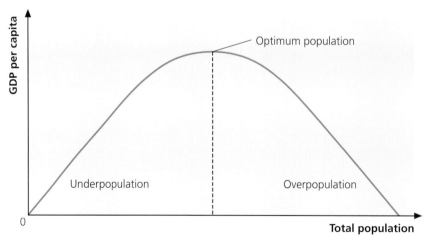

Figure 9.8 The concept of optimum population

In reality, the concepts of optimum, overpopulation and underpopulation are elusive. This is due to several factors:
- incomplete knowledge of any country's resource base

- changes in technology
- environmental changes such as drought reducing the resource base
- difficulties in measuring living standards
- the greater importance of human resources (especially in many developed countries), such as a skilled and educated workforce, compared to natural resources

A further problem arises because the concepts of optimum population, overpopulation and underpopulation are only truly relevant where a society depends on local resources. For example, a society based on subsistence agriculture and experiencing rapid population growth will put increasing pressure on local water, soil and timber resources. The result is likely to be a reduction in crop yields, food shortages and land degradation. Such a society could reasonably be thought of as overpopulated. This situation is found in a number of developing countries, especially in sub-Saharan Africa. In countries like Mali, Niger and Burkina Faso, rapid population growth among nomadic herders and sedentary cultivators in the past 50 years, together with declining water resources due to drought and climate change, has caused overpopulation and desertification.

Governments are often acutely aware of imbalances between population and resources and develop policies to address the problems. These policies can approach the problem of population–resource imbalance by:
- promoting economic development and increasing the resource base (e.g. Brazil)
- controlling migration and/or fertility and influencing the size of the population (e.g. China)

1 Defining rural areas

Rural areas can be defined by criteria such as population size, population density, employment, sociology and land use. There is no internationally agreed definition of rural areas (or urban areas) — each country adopts a scheme which suits local circumstances.

National censuses often identify rural populations by administrative units. These units are determined by population size or population density. Therefore, in England and Wales the rural population are those people who live in rural districts.

Rural societies have distinctive characteristics:
- Compared to urban societies, they are more close-knit, have stronger family ties, attach greater importance to religion, are less mobile and have more traditional lifestyles.
- Extensive land uses, such as agriculture, forestry and recreation and leisure, dominate rural areas.

2 Rural change in the UK

In the last 40 years or so, economic, social and demographic change in rural Britain has been influenced by changes in migration patterns, personal mobility, personal wealth, service provision and planning policies.

2.1 Counter-urbanisation

Counter-urbanisation describes the urban–rural shift of population that has taken place in developed countries since the early 1970s, resulting in an increase in the proportion of rural dwellers. This migration to rural areas is a dramatic turnaround from the urbanisation trend that dominated developed countries in the first half of the twentieth century. Between 1991 and 2000, migration added nearly 840,000 people to the population of rural areas in England and Wales. This trend continued between 2001 and 2006, though at a lower rate of around 35,000 a year (Figure 10.1).

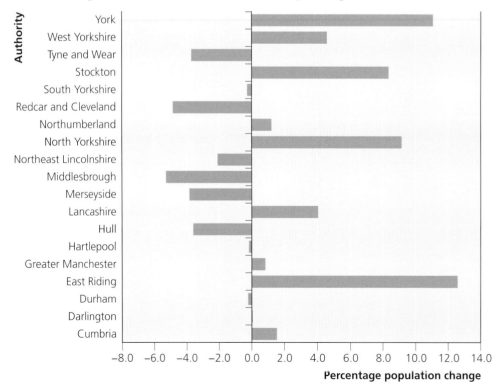

Figure 10.1 Population change in northern England, 1991–2007

Most urban–rural migrants continue to work in the towns and cities they have left. As a result, they need to live within **commuting** distance of their workplace. This explains the rapid population growth in many rural counties and rural districts (e.g. Warwickshire, Cheshire, north Yorkshire) close to major cities and conurbations. It also explains why most migrants are relatively high earners. Only the better-off can afford the extra housing and commuting costs.

However, counter-urbanisation is not just about commuters. Increasing numbers of older people, with generous pensions or valuable properties in urban areas, move to more scenic areas when they retire (e.g. the south coast of England, north Norfolk and the Lake District). In addition, an increasing number of people work from home (using ICT and other electronic systems) and therefore have the freedom to move away from cities.

Both **push factors** and **pull factors** explain the counter-urbanisation process. High crime rates, anti-social behaviour, pollution, traffic congestion, lack of community, a run-down physical environment and poor services are some of the unattractive features of life in large towns and cities. In the UK, the attraction of the countryside (the rural 'idyll') is particularly strong and reflects cultural values that are often quite different from those of other, more urbanised European cultures.

2.2 Rural depopulation

Depopulation is the absolute decrease of the population of an area or place. In the UK, depopulation usually results from net migration loss in rural communities. Some regions, such as the northwest Highlands and islands of Scotland, and central Wales, have suffered depopulation for decades. Lack of services and limited employment opportunities have provided the impetus for out-migration. In lowland regions, declining employment in agriculture has been the driving force behind depopulation. For example, employment in agriculture fell by 70% in Norfolk between 1950 and 1980.

Natural decrease also plays a part in rural depopulation. Most out-migrants are young adults, which reduces future birth rates and increases the average age of the population. Ageing populations mean fewer births and more deaths, increasing the likelihood of natural decrease and depopulation. Rural depopulation has triggered the decline of transport, retailing, educational and medical services in many rural areas in the UK, as Figure 10.2 shows.

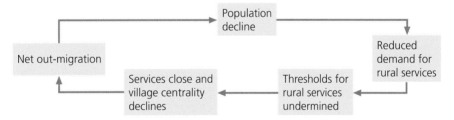

Figure 10.2 Rural depopulation and rural service provision: the vicious circle of decline

Box 1   *Types of rural area*

In the UK, rural areas are classified into two categories: pressured and remote.

■ **Pressured rural areas** are within commuting distance of large urban centres. Most inhabitants have few direct links to the traditional rural economy. Many market towns and villages have expanded to accommodate population growth caused by incoming commuters and their families. New housing estates have appeared, encroaching on the countryside, and barns have been converted to private dwellings. House prices have soared.

■ **Remote rural areas** retain strong links with agriculture and other rural activities. In the uplands, remote rural communities are often among the poorest in the UK. Incomes in hill farming are low and there are few job opportunities for young people. Net out-migration and depopulation have been a feature of these communities for the past 150 years. However, some remote rural areas have experienced a population revival in the past 30 years (e.g. Inverness, Moray and Ross in northern Scotland). The attractions include the quality of life and low cost of living.

3

Declining rural services

Many rural services in developed countries such as retailing, transport, primary schools and healthcare have declined in recent years and are continuing to decline. This raises issues of **inequality of access** to services, not only between urban and rural communities but also between different socioeconomic and demographic groups within rural societies.

3.1 Retailing

The importance of **scale economies** in retailing has led to the dominance of supermarkets and super-stores in retailing during the last 30 years. In the UK in 2008, the four largest supermarkets had a 75% share of the market in grocery spending. This has put pressure on small, independent retailers. Unable to compete with the supermarkets, thousands of small independent shops in rural areas have closed. Two other factors have been important in the decline of rural retailing: the growth of car ownership, giving rural dwellers improved access to supermarkets and other large stores located in nearby towns; and the purchase of fridges and freezers as standard items in most households, which encourages weekly bulk shopping for food. Therefore, in the predominantly rural east of England region in 2008, 49% of parishes were without a shop or retail outlet, only 43% of parishes had access to a post office (compared to 54% in 2000) and barely two-thirds of parishes had a pub (compared to 83% in 2000).

Table 10.1 Percentage change in service provision in England, 2007–08

Service	Rural	Urban
Banks and building societies	−2.6	−1.4
Petrol stations	−1.7	−0.4
Post offices	−5.4	−6.3
GP surgeries	−0.3	+0.1
Primary schools	−0.2	+0.1

3.2 Transport

The growth of car ownership has drastically reduced the demand for rural bus services in the postwar period, as Figure 10.3 shows. In the east of England nearly half of all parishes have no access to national and community buses. Given the decline in rural retail and healthcare services, the withdrawal of rural bus services has a severe impact on those groups — the old and the poor — without access to cars.

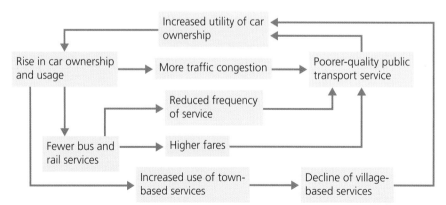

Figure 10.3 The effect of increasing levels of car ownership on public transport services

3.3 Education

Forty years ago, most villages had their own primary school. Today, many educational authorities in the UK have opted to close small village schools and bus children to larger schools in market towns and key villages. There are strong economic and educational arguments against small schools. Small

schools are expensive to run (especially when there are fewer than 50 pupils on the school roll), offer only a restricted curriculum and may be to the social disadvantage of pupils. Against this, village schools provide a focus for rural communities and their closure may leave children facing long journeys to school each day.

Closure is most likely where there are falling school rolls. Falling rolls may have several causes: rural depopulation, the selective out-migration of young adults of child-bearing age from rural communities and the in-migration of older adults without children.

4 Second homes and affordable housing

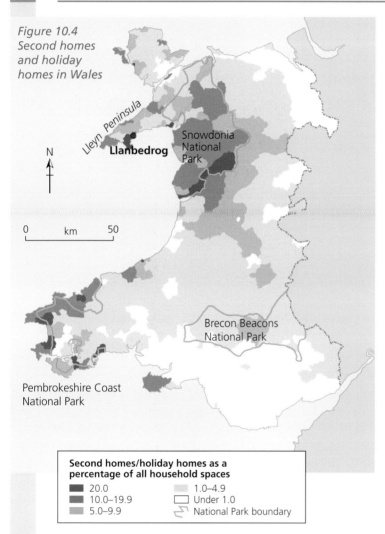

Figure 10.4 Second homes and holiday homes in Wales

Second homes/holiday homes as a percentage of all household spaces

- 20.0
- 10.0–19.9
- 5.0–9.9
- 1.0–4.9
- Under 1.0
- National Park boundary

Second homes are properties that are not someone's main residence and are occupied for less than half the year. In contrast, holiday homes are operated as a business and are rented out to visitors during the holiday season. There are currently around 250,000 second homes and holiday homes in rural areas in the UK. Between 1994 and 2004 the number of second homes and holiday homes increased by nearly 40%.

The recent growth of second homes and holiday homes in the countryside has three main causes:

- a surplus housing stock due to depopulation in some areas
- rising disposable incomes
- increased mobility through car ownership and improved road networks

Geographically, second homes are highly concentrated. In some local authorities up to one-quarter of houses are second homes. Second homeowners favour properties around the coast, in National Parks and in other areas of high environmental quality. Hot spots in the UK include north Norfolk, northwest Wales (Figure 10.4), Devon and Cornwall and the south Lake District.

Second homes and holiday homes are controversial for several reasons:

- occupied only at weekends or holidays, second homes and holiday homes contribute little to the support of local services
- they decrease the local housing supply
- in rural areas where housing is in short supply, second homes push up house prices. This makes housing unaffordable for local people
- they undermine the sense of community and traditions of rural life

In order to curb the growth of second homes in some of the most popular areas, there are suggestions that the current council tax discount could be withdrawn and that legislation could make it illegal to sell an existing permanent residence as a second home.

5 Planning policies in rural areas

5.1 Key settlements

During the past 50 years, the **key settlement policy** has been the cornerstone of rural planning in the UK. The principle behind the policy is the sustainable provision of essential services, jobs and new housing in small towns and larger villages. The key settlements are a focus for development and act as service centres for surrounding rural hinterlands. Also, by concentrating development, key settlements help to reduce commercial pressure on the countryside.

Recent research, however, casts doubt on the importance of the traditional service hierarchy in rural areas. Improvements in personal mobility mean that many rural residents no longer base their lives around the nearest market town. Journeys to work and shopping trips are increasingly to a range of destinations including more distant cities and supermarkets at edge-of-town sites.

5.2 Sustainable management of retail services and affordable homes

In the UK concern about the implications for rural communities of retail service decline and **affordable housing** has produced a range of government-led initiatives (Table 10.2).

Table 10.2 Planning for the sustainable management of rural shops, post offices and pubs

Initiative	Description
Rate relief	Food shops, post offices, pubs, petrol stations and village stores qualify for a mandatory 50% rate relief, which can at the discretion of the council be extended to 100%.
Rural Shops Alliance	National trade association that represents the interests of 7,200 independent village shops and lobbies on their behalf.
Enterprise Development	Encourages and supports local businesses. Low interest loans up to £10,000 available.
Rural Enterprise Network Support	Opposes the closure of rural services and supports the development of independent shops, post offices, pubs, restaurants, garages and cafés.
LEADER	An EU initiative to help rural communities improve their quality of life. Makes funds available for development projects such as farmers' markets.
Sainsbury's Assisting Village Enterprises (SAVE)	An association between village shops and the supermarket giant Sainsbury's. Rural shops are able to stock and sell Sainsbury's products. Around 71% of SAVE stores are rural post offices that sell a small range of convenience goods. Similar partnership schemes have been set up between other large supermarkets and village shops, e.g. Tesco and local village shops in south Norfolk.
Village Shop Grant Scheme	Local authorities such as mid-Suffolk operate rural services schemes to fund and give advice to rural businesses. Grants are available towards new projects. The scheme also extends to community services such as the maintenance of village halls and child care.
Affordable housing	The Affordable Rural Housing Commission reported in 2006 and recommended that 11,000 affordable housing units a year should be built in England in settlements of less than 10,000 people to meet local need. Major new housing developments in rural areas are usually confined to strategic key settlements. Under current government guidelines, developers must ensure that 40% of new dwellings comprise affordable housing.

6 The environmental impact of rural change

In the past 40 years change has affected the rural environment in developed countries as much as rural demography, economy and society. Much of this is due to changes in farming driven by government and, in western Europe, by the European Union (EU). Until recently, these changes have had adverse

environmental effects including pollution, damage to ecosystems and the destruction of habitats, wildlife and historic landscapes. Since 1945 the UK has lost half of its lowland ancient woodland, 250,000 km of hedgerows, 80% of chalk downland, 95% of traditional hay meadows and 80% of wetland fens. Most of these losses were the result of the intensification of farming.

6.1 Arable farming in eastern England

Eastern England, including Norfolk, Suffolk, Essex and Lincolnshire, is the pre-eminent arable area of the UK. Physical conditions, including climate, relief and soil, impose few limits to cultivation, allowing a full range of temperate arable crops to be grown. In the 1970s and 1980s the EU's **Common Agricultural Policy** (CAP) gave arable farmers generous financial incentives (including guaranteed prices) to increase the production of crops such as wheat, barley and oilseed. Farmers responded by increasing the:

- intensity of cultivation, using more agro-chemicals such as nitrates, phosphates and pesticides
- area under cultivation by removing hedgerows, clearing patches of ancient woodland and draining wetlands

Meanwhile, farming became organised on an increasingly large scale. Farms were amalgamated, creating huge **agribusiness** enterprises, which lowered costs through economies of scale. Production became ever-more **automated**.

6.2 Intensive farming and loss of biodiversity and habitat

Farm amalgamation, farm **economies of scale** and increasing **arable specialisation** in the UK since 1945 have resulted in the loss of almost 250,000 km of hedgerows (Figure 10.5). Hedgerow removal was an essential step in the automation of farming, creating larger fields and allowing the use of bigger farm machines. Removing hedgerows also increased the land available for cultivation.

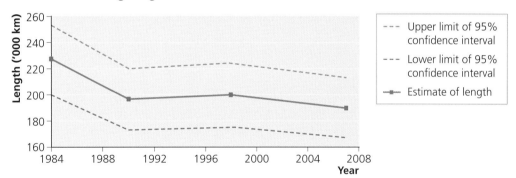

Figure 10.5 Hedgerow length, eastern England

However, the environmental impact of these changes was ruinous. For example, in Norfolk and Suffolk, historic rural landscapes of small fields, hedgerows and mature trees on the clay lands were transformed into bleak, windswept prairies. Hedgerows are an important habitat for wildlife and a valuable ecological resource. They provide shelter, food, breeding and nesting sites and corridors for the movement of wildlife. They also provide a refuge for wild plants unable to survive in the intensely cultivated arable fields. Their destruction not only caused a loss of **biodiversity**, it also fragmented existing wildlife habitats.

The intensification and specialisation of arable farming had other environmental disadvantages. Ancient lowland woodlands have been reduced by half since 1945, and wetland fens and mires by 80% in the same period. Meanwhile, chemical **fertilisers** and **pesticides** have been a major source of water pollution in arable areas of eastern England. **Leaching** removes nitrate and phosphate fertilisers from the soil, polluting groundwater, surface streams and lakes. Nutrient enrichment of surface waters leads to algal growth and **eutrophication**, which kills fish and aquatic invertebrates and pollutes public water supplies

6.3 Managing farming change and the environment

The CAP and environmental policies

In the past 20 years the CAP has been reformed to make modern agriculture in the EU environmentally sustainable. The sustainable development of rural areas has become one of the CAP's two central 'pillars'.

The old system of **subsidies** and guaranteed prices, which gave incentives to farm intensively, has been replaced by the **Single Farm Payment Scheme (SFPS)**, which pays producers for each hectare farmed. This is good for the environment. It reduces over-production and encourages farmers to grow crops that the market wants. To get the payments there are **cross-compliance** rules which benefit the environment, such as repairing hedgerows, reducing inputs of pesticides, using less chemical fertiliser and leaving strips of uncultivated land around field margins. Where farmers deliver additional environmental services, such as preserving hedgerows, they can receive extra payments.

Set-aside

The **set-aside** scheme, introduced in 1992, allowed arable farmers to remove a certain proportion of their land from production in return for a fixed payment per hectare. In 2007, 8 % of England's arable area was in set-aside. On set-aside land, farmers must:

- grow a green cover crop such as grass or allow natural vegetation to establish itself
- leave the land untouched between January and mid-July
- not use any agro-chemicals

Set-aside has been hugely beneficial to wildlife. Cover crops provide food and habitats for insects and birds and as the land is not ploughed until mid-July ground-nesting birds are undisturbed. However, in 2008, in order to offset poor harvests in 2007 and soaring food prices, the EU decided to abolish set-aside. An alternative has yet to be agreed. Two possibilities are:

- farmers will compulsorily have to manage a small percentage of their cultivated land for environmental purposes
- farming practices will be encouraged voluntarily to protect wildlife habitats and retain and manage uncropped land

Nitrate pollution

Intensive farming is a major source of water pollution by nitrates, phosphates, pesticides, silage and manure. Since the Nitrates Directive of 1991, farmers in many parts of England have received financial compensation for applying smaller amounts of nitrate fertiliser. Today, as part of cross-compliance, farmers have to limit both nitrate and manure applications.

Table 10.3 The UK government's agri-environmental schemes in England

Initiative	Description
Environmentally Sensitive Areas (ESAs)	There are 22 ESAs in England, which cover around 10% of the agricultural area — nearly 6,000 km². The scheme, which is now closed to new areas, has been succeeded by the Environmental Stewardship Scheme (ESS). Farmers in ESAs enter an agreement to farm in an environmentally friendly way. In return, they receive grants to compensate them for loss of income. The aim is to safeguard and enhance parts of the country with high landscape, wildlife and historic value.
Environmental Stewardship Scheme (ESS)	This scheme was established in 2005 and replaced earlier ESAs and the Countryside Stewardship Scheme. It rewards farmers for conservation and environmental enhancement of the countryside. Natural England is responsible for the ESS programme. Under the scheme, farmers and other land managers get funding for 'effective environmental management'. This can mean using farming that conserves wildlife and biodiversity, protects historic environments and natural resources and maintains or enhances landscape quality.
Farm Woodland Premium Scheme (FWPS)	Farmers entering this scheme are paid to convert agricultural land to woodland. The environmental gains are new habitats for wildlife, greater biodiversity and improvements in the quality and character of the countryside. Annual payments compensate farmers for loss of agricultural income. Payments are made for 10 years for conifer woodlands and for 15 years for deciduous woodlands.

TOPIC 11 Urban change: problems and planning

1 Defining urban populations and urban areas

Classifications of urban populations and urban settlements depend on census definitions. Population size is most often used to define urban areas, but thresholds vary between countries. In several countries in Latin America and West Africa, the threshold population is 2,000, whereas it is 200 in Iceland and 10,000 in countries such as Italy and Benin.

Apart from population size, other criteria often used are population density, distance between buildings within a settlement, economic activity and legal or administrative urban boundaries.

Determining the areal extent of towns and cities poses similar problems. The size of any town or city depends on the boundaries chosen. There are several possibilities and each is likely to give a very different population estimate:

- legal or administrative boundaries
- the contiguous built-up area (including both the inner and outer suburbs)
- the contiguous built-up area and the physically separate **exurbs**
- the contiguous built-up area and the surrounding commuter **hinterland**

2 Urbanisation and urban growth

Urbanisation is the proportion of urban dwellers in a country or region. **Urban growth** describes the increase in the number of urban dwellers. It is clear that when levels of urbanisation increase, there must be a relative shift of population from rural to urban areas. Such a shift often occurs when:

- rural–urban migration produces a net migrational gain in urban areas and a net migrational loss in rural areas
- rates of natural increase are greater in towns and cities than in the countryside

3 Global urbanisation

3.1 Urbanisation by continent

At the global scale, rapid urbanisation has occurred in the past 50 years (Figure 11.1). By 2010 half the world's population lived in towns and cities. Europe, North America, South America and Oceania are the most urbanised continents; Africa and Asia are least urbanised. Although only 43% of Asia's population lives in towns and cities, the absolute number of urban dwellers in Asia (around 1.7 billion) is by far the largest of any continent (Figure 11.2).

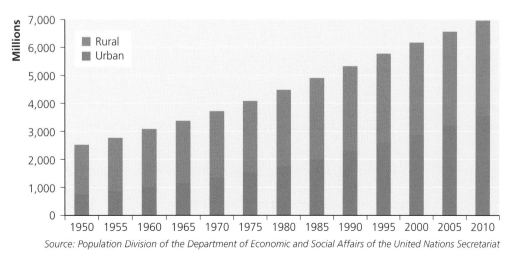

Source: Population Division of the Department of Economic and Social Affairs of the United Nations Secretariat

Figure 11.1 Global change in urban and rural populations, 1950–2009

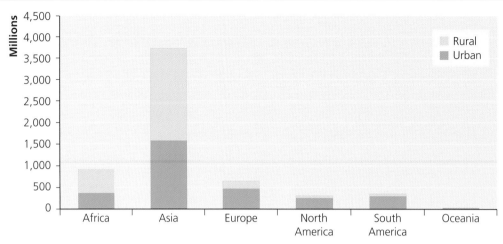

Figure 11.2 Urbanisation by region, 2010

Today, urbanisation is concentrated in the developing world, especially in Africa and Asia (Figure 11.3). This trend will continue so that by 2025 more than half the population in Africa and Asia will live in urban areas, and four out of every five urban dwellers will be in developing countries. In Europe, North America and Oceania, urbanisation levels peaked in the mid-twentieth century and have fallen steadily in the last 30 years.

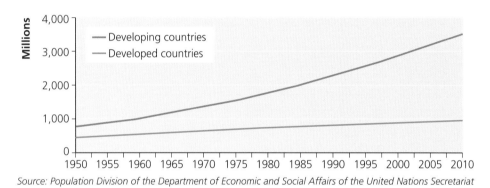

Source: Population Division of the Department of Economic and Social Affairs of the United Nations Secretariat

Figure 11.3 Urban population growth in developed countries and developing countries, 1950–2010

3.2 Demographic causes of urbanisation

Urbanisation in developing countries results from both natural increase and rural–urban migration. On average, natural increase accounts for around 60% of annual urban population growth in developing countries; the remaining 40% is due to rural–urban migration. Where migration makes a major contribution to urban population growth, annual rates of increase are often very high. Some of the fastest growing cities are in China. Cities such as Chongqing and Shenzhen grew by more than 10% a year between 2000 and 2005.

4 *Urban growth and city size*

Rapid urban growth in the past 50 years has had two important consequences for city size:

- There has been a huge increase in large cities with populations in excess of 1 million. In the early 1960s there were 113 so-called 'million cities', in 1990 there were 281 and by 2009 there were 433.
- The growth of these 'million cities' has been mainly concentrated in the developing world. The largest cities, with a population of more than 10 million, are known as **mega cities** (Table 11.1). Like million cities, the number of mega cities will increase rapidly in the next 25 years and, again, the increase will be mainly concentrated in the developing world.

Table 11.1 The world's largest cities, 2006

Rank	City/urban area	Country	Population in 2006 (millions)
1	Tokyo	Japan	35.53
2	Mexico City	Mexico	19.24
3	Mumbai	India	18.84
4	New York	USA	18.65
5	São Paulo	Brazil	18.61
6	Delhi	India	16.00
7	Kolkata	India	14.57
8	Jakarta	Indonesia	13.67
9	Buenos Aires	Argentina	13.52
10	Dhaka	Bangladesh	13.09
11	Shanghai	China	12.63
12	Los Angeles	USA	12.22
13	Karachi	Pakistan	12.20
14	Lagos	Nigeria	11.70
15	Rio de Janeiro	Brazil	11.62
16	Osaka–Kobe	Japan	11.32
17	Cairo	Egypt	11.29
18	Beijing	China	10.85
19	Moscow	Russia	10.82
20	Metro Manila	Philippines	10.80
21	Istanbul	Turkey	10.00

5 *World cities*

The world economy has become increasingly globalised in the past three decades. Leading this trend are large **transnational corporations** (TNCs) with factories, offices and subcontractors in many countries. Today, they dominate the global economy and international trade. Underpinning **globalisation** is a hierarchy of world cities that operate as the command and control centres for the global economy.

Three cities sit at the top of the global urban hierarchy: New York, London and Tokyo. They are the headquarters locations for many TNCs and centres of world finance. They provide international **producer services** in areas such as accounting, law, advertising and consultancy. Their importance has made them global hubs in international communications.

Below the truly global cities are three lower orders of world cities:
- cities linked to large international (but not global) areas, such as Los Angeles (Pacific Rim) and Singapore (southeast Asia)
- cities that link large national economies with the global system, such as Paris and São Paulo
- cities that integrate important sub-regional economies with the global systems, such as Seattle–Vancouver (Pacific Northwest) and Osaka–Kobe (Kansai)

TOPIC 11 Urban change: problems and planning

6 Urban social and economic change in developed countries

In developed countries, urban change has given rise to social and economic issues that affect the CBD, inner city, suburbs and the rural–urban fringe

6.1 Central business district

In the USA, the CBD has lost much of its importance as the pre-eminent centre for retailing activities. For examples, Los Angeles' city centre is dominated by commercial offices, leisure-based activities such as restaurants and cafés, public buildings and local authority administration. Today, retailing is concentrated in huge, free-standing shopping malls in the suburbs. Location in the suburbs gives better access to customers as well as providing extensive parking lots, vital in a car-based society like the USA. A similar trend has occurred in British cities in the past 30 years but on a much smaller scale. Governments have restricted the decentralisation of retailing, arguing that the loss of retailing triggers irreversible city-centre decline. Therefore, planners, guided by government policy, have been reluctant to approve new suburban shopping malls and retail parks on the American model. As a result, the CBDs in most British cities have largely retained their status within the intra-urban shopping hierarchy.

Meanwhile, investment in city centre shopping malls, pedestrianisation and the growth of leisure retailing (e.g. clubs, bars, restaurants, cafés and hotels) have reinvigorated and broadened the appeal of British city centres. In addition to employment and retailing, city centres such as Newcastle, Manchester and Leeds have developed a distinctive culture, with nightlife that attracts tourists and other visitors as well as local people. After decades of decline, people have begun to move back into British city centres. Thousands of new apartments have been built and old office blocks, mills and warehouses have been converted for residential use. In 1990 only 1,000 people lived in Manchester's city centre. By 2005 this figure had risen to 20,000. Local authorities have encouraged this movement, which has brought in investment, revitalised run-down areas and has created a vibrant city centre. It also reduces journeys to work and is therefore consistent with the idea of building **sustainable cities**. Recentralisation has been led by young couples and singles (half of all apartments in central Manchester are single occupancy). The attractions are lifestyle, leisure and cultural facilities, nightlife and proximity to work in the centre.

6.2 Inner city

Socioeconomic issues in British inner cities focus on **inequality**, **multiple deprivation** and **social exclusion**. Inner-city populations are often poorly qualified, with low incomes and high rates of youth unemployment. Ethnic minorities, attracted by low-cost housing and cultural preferences, are disproportionately represented in inner-city areas. Large concentrations of ethnic groups such as those in Brick Lane in London, St Pauls in Bristol and Manningham in Bradford form **ghettoes**, with strong spatial segregation and poor integration with the host society.

Box 1 *Multiple deprivation*

In the UK the **multiple deprivation index** (MDI) is used as an objective measure of poverty. The index, made up of seven components — income, unemployment, health, education, skills, housing and crime — ranges from 1 (least deprived) to 100 (most deprived) (Figure 11.4). MDI scores are published as tables for England's 8,184 wards and parishes and at a smaller scale for the 32,482 census super output areas. A number of problems are connected to poverty, including poor housing, ill health, poor education and skills and crime. However, these problems are not exclusive to inner-city areas. They also feature in many local authority housing estates in the suburbs of major British cities.

Figure 11.4 Multiple deprivation index

CASE STUDY 1	Harehills, Leeds: inner-city deprivation and decay
Background	Harehills is an inner-city suburb in east Leeds, 1 to 2 km from the city centre (Figure 11.5). The suburb was added between 1870 and 1914. The original housing (most of which survives today) comprises small terraces, much of it back-to-back. Two-thirds of the housing is rented from the council, social landlords (e.g. housing associations) and private landlords. Low-cost housing has attracted low income groups, with large ethnic minorities principally of African-Caribbean and south Asian origin. One quarter of the ward's population was born outside the UK.
Social and economic issues	A large proportion of Harehills' residents suffer multiple deprivation. In terms of the MDI, Harehills is among the poorest 5% of wards in England. Unemployment is twice the Leeds average. High levels of unemployment are related to poor levels of education and skills. Some 45% of the adult population have no educational qualifications and household incomes are less than half those of the more prosperous wards in the city. Life expectancy is 4 years less than the average for Leeds and at the 2001 census 12% of residents reported poor health. Sub-standard housing and overcrowding also contribute to poor health. Reported crime is four times the average for Leeds, with particularly high levels of domestic burglary, vehicle crime and criminal damage.
Responses	Harehills is included in the East and Southeast Leeds (EASEL) Regeneration Project. This is a £1 billion partnership programme between Leeds City Council, developers, employers and various agencies. Over 20 years, it aims to revitalise this part of Leeds. In Harehills the emphasis is on housing improvements including demolition of the worst housing and upgrading the environment. Healthcare will be improved to tackle high rates of ill health and mortality. This will include better primary healthcare facilities and a new children's unit in the local hospital. Retail provision will be improved by renovating the Harehills Corner shopping centre. The most disadvantaged areas in Harehills (among the 3% most deprived areas in the country) will also benefit from the Neighbourhood Renewal Fund, which in 2007–08 was worth nearly £15 million. Its aims are to improve the quality of life and employment prospects of residents.

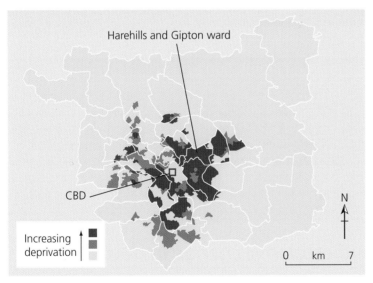

Figure 11.5
Location of Harehills

Many inner-city suburbs in the USA have been transformed by **gentrification**. Gentrification describes the process by which young, higher-income groups move into run-down inner-city areas. As a result, these areas become socially, economically and environmentally upgraded. Examples include San Francisco's Mission District, Dorchester in Boston and Chelsea/Clinton in New York. However, gentrification is controversial because it often drives up rents and forces out existing low-income residents. The winners appear to be the better-off; the losers are the poor.

6.3 Suburbs

The suburbs of most British cities grew rapidly during the inter-war and post-war years. Growth was initially driven by improvements in public transport (e.g. trams, suburban trains, bus services) and later by private car ownership. In contrast to the inner city, suburban housing was lower density — typically

detached and semi-detached dwellings with gardens. Open areas such as playing fields and parks studded the suburbs, add to the feeling of space.

Social, economic and demographic changes have created problems in the suburbs. Many local authority estates suffer worse deprivation than inner-city neighbourhoods. Unemployment, ill health, poor schools, poor skill levels and crime are often endemic. A lack of shops selling fresh food contributes to poor diets and ill health. Unlike inner-city areas, local authority estates are occupied almost exclusively by low-income white communities (e.g. Pennywell in Sunderland, Blackbird Leys in Oxford, Bransholme in Hull). Social dysfunction has been amplified as more aspirational residents have moved out to become owner-occupiers and as local authorities have housed 'problem' families on these estates.

In the outer suburbs, where the built area meets the countryside, there are problems of **urban sprawl**. The problem is likely to get worse. Natural population growth, immigration and the increase in single person households in the UK mean that the demand for new homes is forecast to grow by nearly 225,000 a year until the mid-2020s. Demand is particularly high in England and in the southeast region, where 654,000 new homes are planned by 2026. Government targets are that at least 60% of new homes should be built on previously developed (**brownfield**) land. This puts huge pressure on **green belt** land (Figure 11.6), raising issues of loss of countryside and amenity, and the sustainability of urban growth.

Figure 11.6 Green belt land in England

CASE STUDY 2   Threats to the green belt in Surrey

The South East Plan (SEP), published in May 2009, determined where 52,480 new homes should be located across the so-called London Fringe authorities by 2026. Although the plan aims to meet housing needs mainly within urban areas, it is clear that encroachment on existing green belt land will be inevitable. Local people in Guildford, Reigate, Banstead and other Surrey towns strongly oppose any threat to the green belt. Support has come from Friends of the Earth and the Campaign to Protect Rural England. They argue that development of green belt land will raise carbon dioxide emissions, increase traffic congestion, place huge pressure on water resources and destroy countryside. Even the SEP concedes that the scale of new housing planned in the southeast will have detrimental effects on the environment, including poorer air quality, loss of biodiversity and increased flooding due to development in flood-risk areas.

7   Urban social and economic change in developing countries

Rapid urbanisation and population growth in the past 50 years have resulted in huge social and economic problems in cities in developing countries. These problems include housing shortages and poor housing quality, lack of employment and inadequate service provision.

7.1 Housing

Urban populations in developing countries are increasing rapidly and most urban dwellers are poor. For example, Nairobi's population increases by 200,000 every year and 60% of its inhabitants live in **slum settlements** on just 5% of the city's land area.

Today, most urban authorities in developing countries acknowledge that self-help, through the construction of **informal settlements**, is the only practical solution to the housing problem. These settlements

are often illegal, unplanned and insanitary. Initially, they lack any service infrastructure such as mains water supply, sewerage, drainage and electricity. Houses are improvised shacks made of wood, brick, mud, corrugated iron and other rudimentary materials.

In southern and east Africa, many squatters are tenants who pay rent to slum landlords. Rents are often high (an estimated 15% of household income in Kibera slum in Nairobi, see Table 11.2) and contribute to poverty. Population densities in some settlements exceed 300,000 persons/km^2 and overcrowding contributes to ill health and the spread of infectious diseases such as tuberculosis.

In the past, many urban authorities adopted a hard line towards residents, regularly evicting them and destroying informal settlements. They were regarded as hotbeds of crime, disease and political dissent. Even where they were tolerated, because of their illegal status no provision was made to upgrade basic services. Now attitudes have changed. Informal settlements are seen as a way forward. Once granted legal title to the land, the residents gradually improve their homes. Eventually, former slum settlements are upgraded and absorbed into the city to become valuable additions to the city's housing stock.

Table 11.2 Master plan to upgrade Kibera, Nairobi

Feature	Description
Physical infrastructure	Provide effective sanitation, water supply, access roads, storm water drainage, electricity and street lighting.
Social infrastructure	Provide schools, health centres, community centres, orphanages, recreational facilities and open spaces.
Housing	Give residents security of tenure. Build low-cost housing and cooperative housing. Provide resources for housing improvements.
Environment	Organise rubbish collection and treatment and clean up the badly polluted River Ngong that flows through Kibera.
Employment	Establish markets, shops, kiosks and shopping centres. Introduce micro-finance and credit schemes.
HIV/AIDS	Establish education awareness programmes, counselling, test centres and HIV clinics.

7.2 Improving informal settlements

Most cities in the developing world have low-cost housing schemes for the poor. For the urban poor, low-cost housing is often a less attractive option than self-built housing in informal settlements. There are a number of reasons for this:
- informal settlements are often more conveniently located in relation to employment
- most residents either pay low rents or live rent-free
- residents have workshops and other sources of livelihood in or near their homes
- residents have strong social ties in informal settlements
- residents may have sufficient political power to force the city authorities to upgrade their settlements
- if residents are granted legal title to the land, many will get ownship of their homes

In Delhi, 2.4 million people live in informal settlements known locally as ***jhuggies***. Most *jhuggies* have few amenities and services. Infant mortality in such settlements is twice the average for India. The city authorities hope to improve living conditions by helping the residents to help themselves. They grant squatters legal title to their land, provide essential infrastructure such as mains water, roads and electricity and give loans to residents to purchase building materials and hire builders to improve their homes.

Sites-and-services schemes are a popular, low-cost solution to the housing problem. The city authorities provide serviced plots with mains water, electricity, sewerage and roads. Residents build their own houses on the plots, either from scratch or around a basic shell.

7.3 Employment

Growth rates in manufacturing and service industries in cities in developing countries have failed to keep up with population growth. Only a small proportion of workers get jobs in the modern or **formal sector**. However, without any welfare support from the state, people cannot afford to be unemployed. As a result, people must rely on self-help and create their own jobs.

This process has spawned a huge **informal sector**. The sector is labour-intensive and 'soaks up' millions of workers who otherwise would be unemployed. It consists of small enterprises that often rely on traditional technology and skills. Much of the work in this sector is home-based and casual. The informal sector is unregulated and untaxed and workers are not protected by employment law.

Activities in the informal sector are extremely diverse. They include producers such as builders and various workshop manufacturers, retailers such as street hawkers and service providers such as shoe cleaners and launderers. Some activities, such as drug dealing, smuggling, theft and prostitution, are criminal or socially undesirable. Even so, the informal sector is vital to the economic functioning of cities in developing countries (in 2001 the informal sector accounted for one-quarter of Kenya's non-agricultural GDP and three-quarters of national employment) and the survival of millions of their inhabitants.

8 *Urban inequality*

Social and economic inequalities are found in all countries, although they tend to be more extreme in developing countries. Some of the widest inequalities are in sub-Saharan Africa, the world's poorest region. In contrast, the world's richest countries, such as the social democracies of Scandinavia, support the most equitable societies. Inequality often relates to particular groups in society, defined by race, ethnicity, migration, gender and age. These groups may suffer **multiple deprivation** such as low incomes, high unemployment, poor access to services, high rates of crime, ill health and environmental pollution. Although inequalities exist at all scales, they are most conspicuous within urban areas. In cities in developing countries it is not unusual for gated communities and high-rise apartment blocks, where affluent families live, to be situated within a few hundred metres of squatter camps and shanty towns.

In geographical terms, urban inequality is expressed through spatial segregation. In developed countries wealthier groups have traditionally been located in the outer suburbs, with poorer groups closer to the centre. Indeed, the whole process of **suburbanisation** is about the better-off moving away from central to peripheral locations. Such movements, best exemplified in US cities in the twentieth century, were motivated by a desire on the part of middle and upper income groups to:
- access the social and environmental benefits of the suburbs
- distance themselves from lower income groups (often in-migrants), often perceived as socially inferior

However, there were exceptions to this trend. In some cities, high income groups remained in central areas such as Beacon Hill in Boston and Knob Hill in San Francisco.

In the UK the location of poverty within cities reflects two factors:
- Poorer groups are constrained to purchase or rent low-cost housing. Therefore, the distribution of low-cost rent and social housing provides a basic template, forcing poorer groups to live either in the inner city or in peripheral social housing estates.
- Some ethnic groups are significantly less prosperous than the host community. This is especially true of more recent immigrants (e.g. from south Asia, southwest Asia and east Africa). However, even long-established ethnic communities from Pakistan, Bangladesh and the Caribbean are more likely to be poor than the host population.

In cities in developing countries wealthier groups often live close to the city centre, where there is good access to employment and services. The poorest groups often occupy peripheral locations in informal

settlements, where vacant land for squatting is more readily available. Service provision is often poor in these locations and employment opportunities in more central areas may involve long commutes every day.

9 Urban areas and the environment

9.1 Air pollution

Rapidly expanding urban populations in developing countries, coupled with rising material aspirations, have contributed to pollution and widespread **environmental degradation**. In some cities, especially in emerging economies such as China and India, pollution is a major threat to human health. However, urban pollution is not confined to the developing world. Table 11.3 shows that levels of nitrogen oxide pollution exceed the World Health Organization (WHO) minimum guidelines in four of the world's wealthiest cities (New York, Los Angeles, London and Paris).

Table 11.3 Air pollution in cities

City	Particulates (micrograms/m³)	Sulphur dioxide (micrograms/m³)	Nitrogen oxide (micrograms/m³)
Chongqing	123	340	70
Delhi	150	24	41
Johannesburg	33	19	31
London	21	25	77
Los Angeles	34	9	74
Mexico City	51	74	130
New York	21	26	79
Paris	11	14	57
Santiago de Chile	61	29	81
Taiyuan	51	74	130
WHO guidelines (maximum concentrations)	*20*	*20*	*40*

Source: World Bank

Massive urban growth in China since 1979 has created huge environmental challenges, putting pressure on air, land, water, wildlife and other resources. Urban air pollution is of most concern. In 2007 the World Bank reported that 16 of the world's 20 most polluted cities were in China. Severe air pollution is caused by coal-fired power plants and large concentrations of heavy industries like steel-making, coke-making and chemicals. Particulate and sulphur dioxide emissions from these sources are responsible for smog, **acid rain** and high rates of respiratory illness (e.g. heart disease and lung cancer).

Urban air pollution from motor vehicles is another area for concern. Beijing already has 3.6 million motor vehicles. Motor vehicles emit nitrogen oxide and ozone (both injurious to human health), which form brown **photo-chemical smogs** especially during the summer months. Meanwhile, pollution of rivers and lakes by industrial effluent and untreated sewage adds to environmental damage, contaminates drinking water and is a further threat to human health.

9.2 Solid waste

A major environmental problem is the disposal of solid waste generated by large urban areas. London generates 4.4 million tonnes of municipal household waste a year. Most of this waste is disposed of in 18 landfill sites. Some of these sites are situated more than 120 km from the capital. Landfill is an environmentally unsatisfactory method of waste disposal for three reasons:

- leakage of toxic chemicals from the site
- emissions of methane, a potent greenhouse gas
- the loss of countryside and amenity, and of brownfield sites (e.g. old quarries) that could be used for other purposes in the crowded southeast

London is rapidly running out of landfill space and needs to find an alternative in the next few years. One approach is to extend recycling schemes and simply reduce the amount of waste going to landfill. However, to achieve the city's current targets will require Londoners to expand their recycling sixfold. At the moment London is not investing enough in recycling to meet this target.

9.3 Traffic congestion

Traffic congestion imposes both economic and environmental costs. The UK, with 34 million motor vehicles in 2009, has some of the most congested roads in the world. Delays caused by congestion cost the country an estimated £30 billion a year. Figures 11.7 and 11.8 show how, as the number of cars on the roads has increased, journey times have slowed. Table 11.4 outlines some of the schemes that have been set up to alleviate congestion on the UK's roads.

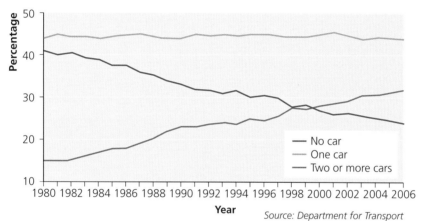

Figure 11.7 UK car ownership, 1980–2006

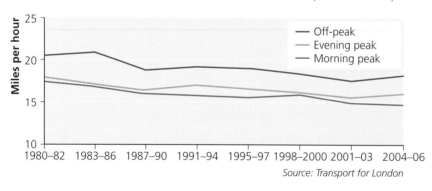

Figure 11.8 Average traffic speeds in London, 1980–2006

The environmental costs of traffic are also high. In Santiago, the capital of Chile, public transport is based on hundreds of old diesel buses. As well as causing congestion, these add significantly to air pollution. One effect is to raise nitrogen oxide levels to twice the WHO limit. High levels of airborne pollutants affect the health of the city's inhabitants, raising morbidity (ill health) and mortality levels. Annual costs of traffic-generated pollution exceed US$500 million. Particulates from diesel engines (three times the WHO limits) cause respiratory problems and are a specific threat to children.

The Urban Transport Plan for Santiago aims to reduce traffic congestion and pollution by:
- reducing average journey lengths
- improving air quality through reduced emissions of sulphur dioxide and nitrogen dioxide
- improving access to public transport
- improving mobility

Table 11.4 Management responses to urban traffic congestion in the UK

Scheme	Description
Congestion charge	Introduced in London in 2003. It initially covered the central area and was extended to parts of west London in 2007. Motorists pay £8 a day to enter the congestion zone.
Light rail, trams etc.	Several large cities such as Sheffield, Nottingham and Manchester have built light rail and tram systems in the past 20 years.
Park-and-ride schemes	Large parking lots on the edge of town linked to the city centre by frequent shuttle bus services. Schemes are in place in cities such as Norwich, Oxford and Durham.
Bus lanes	Dedicated bus lanes, usually in operation during morning and evening rush hours.
Car sharing	Lanes on motorways reserved for cars with two or more occupants.
Integrated transport schemes	Integration of bus and rail services at key interchanges. Synchronised timetables for bus and rail.

To achieve these aims involves investments to expand the metro system and suburban rail services, introducing segregated bus lanes, converting taxis to run on compressed natural gas fuel, reforming the public bus system and developing road pricing.

Investment in public transport, and especially rapid transit, tram and light rail systems, is an important step to reducing traffic congestion and increasing sustainability. In the USA, cities such as New York and San Francisco have successful rapid transit networks. In contrast, Los Angeles has relied heavily on its motorways and private cars. It began to invest in rapid transit only recently (Table 11.5).

Table 11.5 Rapid transit and light rail systems in Los Angeles, New York and San Francisco, 2007

City	Network (km)	Stations	Weekday passengers
Los Angeles	118	62	300,000
New York	369	468	5,000,000
San Francisco	167	43	375,000

Sustainable cities

10.1 Sustainability and urban areas

Sustainable development is 'development that meets the needs of the present without compromising the ability of future generations to meet their own needs'. It gained credibility among planners and politicians following the UN Earth Summit conference in Rio de Janeiro in 1992. The concept of sustainable cities has relevance today because the combination of urbanisation and the rising material aspirations of billions of people in the developing world is having an adverse impact on society and the environment. It is, however, debatable whether urban growth can ever be truly sustainable. For the foreseeable future, cities are likely to remain:
- net consumers of land, energy, water, food and other resources
- dependent on surrounding natural systems to absorb their waste products

Sustainability in an urban context generally means slowing rates of development and living within ecological limits.

10.2 Ecological footprint

The concept of the **ecological footprint** is 'the amount of productive land and water a given population requires to support the resources it consumes and the absorption of its wastes'. The smaller the ecological footprint of a city or a society, the closer it comes to **sustainability**.

Ecological footprints are measured in units of global hectares (gha). One gha has an annual biological productivity equal to the world average. The global biosphere has total of 11.2 billion gha, which averages out at 1.8 gha for every person on the planet. However, actual usage is 2.2 gha per person, indicating an annual drawdown of natural capital. This tells us that current resource use is **unsustainable**.

When applied to cities, the ecological footprint reveals how far urban populations are from getting anywhere near sustainability. The geographic footprint of a city bears no relationship to its ecological footprint. London, for example, needs 49 million gha to produce and dispose of all the energy and materials needed to support its 7.5 million inhabitants. This averages out at 6.63 gha per person, which is three times the global average. The ten cities in England with the largest ecological footprints are shown in Figure 11.9.

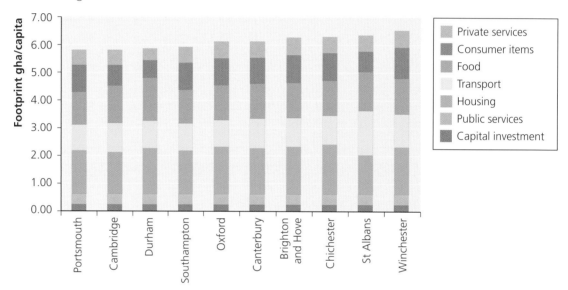

Figure 11.9 Ecological footprints in England

10.3 Eco-cities and eco-towns

China is leading the way in planning the world's first fully sustainable **eco-cities**. Dongtan, which is located on the alluvial Chongming Island at the mouth of the Yangtze River, provides the model for future Chinese eco-cities.

China's planned eco-cities have a number of features. They aim to be **carbon neutral** and will rely on renewable energy such as wind and solar power. They will collect and recycle rainwater and use green building technologies such as grass roofs. The built environment will include extensive green spaces, which will promote evaporation and cooling in summer. Houses will be well insulated and energy efficient and residents will be able to walk or cycle rather than make journeys by car. Public transport will replace private car ownership.

In 2008 the UK government invited bids from developers to build 15 **eco-towns** (Figure 11.10). Although less ambitious than China's eco-cities, the UK's eco-towns will have zero carbon emissions. The government also promised that:

● the eco-towns would not be built on green belt land
● 30% of new housing would be affordable homes
● significant use would be made of brownfield land
● the towns would be car free

Typical of the sites under investigation is Rossington in South Yorkshire, a former mining village with extensive areas of brownfield land, where 15,000 new homes are planned.

Leeds City Region
Sites sought in former
coal-mining district of Selby

Rossington
Up to 15,000 homes,
brownfield site

Rushcliffe
Near Nottingham

Curborough
5,500 homes on brownfield site

Middle Quinton
6,000 homes on old army depot

Weston Otmoor
10,000–15,000 homes, many
low cost. Loss of wildlife

Bordon-Whitehill
5,500 homes (2,000 low cost)
on MoD land. Loss of wildlife

St Austell
5,000 homes on former
china clay mines

Pennbury
Up to 15,000 homes (4,000
low cost) near Leicester

Manby
5,000 homes for people
displaced by coastal erosion

Marston
Two schemes. Up to
30,000 homes. Loss of
farmland and villages

Coltishall
5,000 homes (2,000 low cost)
on former airfield

Hanley Grange
8,000 homes (3,000 low cost)
on greenfield site

Northeast Elsenham
Up to 5,000 homes (1,800
low cost) on greenfield site

Ford
5,000 homes, partly on
old airfield

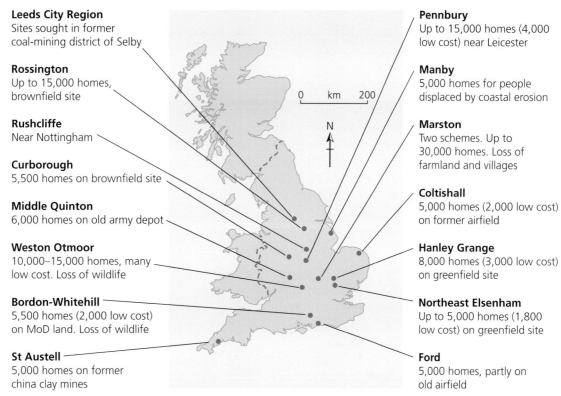

Figure 11.10 Planned eco-towns

TOPIC 12 Globalisation

1 What is globalisation?

Globalisation usually refers to an economic process that involves increases in the flows of ideas, people, goods, services and capital between countries and on a worldwide scale. In the past 30 years or so these flows have intensified, resulting in a much greater level of integration in the world economy and closer economic ties between countries. Globalisation has increased world production of goods and services, expanded world trade and brought prosperity to millions. However, its impact has been geographically uneven. Many millions of people, especially in the poorest developing countries, have received little benefit. Moreover, as the recent world recession has demonstrated, a more integrated global economy does have a downside. What started as a crisis in the USA's housing market quickly spread to banking, financial services and manufacturing. Not only was the USA's economy badly hit, but the entire global economy entered its deepest recession since 1945.

Globalisation has resulted in:

- a greater proportion of manufactured goods being produced by transnational corporations
- large increases in the volume of international trade
- increases in flows of raw materials, as part of the international supply chain
- offshoring of services such as ICT, billing and customer care from developed countries to countries in the developing world
- greater dependence on global flows of capital, controlled by major world cities such as New York, London and Tokyo
- freer international movement of goods, capital and people

Globalisation also has a cultural dimension. Companies such as McDonald's and Starbucks transmit American consumer values and lifestyles. People in developing countries often aspire to Western consumerism, which undermines traditional values and creates more homogeneous societies.

2 The causes of globalisation

The globalisation phenomenon has been driven by five major factors (Figure 12.1): the liberalisation of trade and finance; transport improvements; communications improvements; labour costs and skills; and proximity to markets.

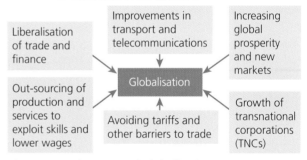

Figure 12.1 The cause of globalisation

2.1 Liberalisation of trade and finance

The World Trade Organization (WTO) provides a forum for discussion on international trade and the reduction of trade barriers. In 2009 it comprised 153 member states. The WTO has played a leading part in promoting **free trade** between member countries, which has stimulated an expansion of world trade and globalisation. Controls have also been removed on international movements of capital and services, giving globalisation even greater impetus.

2.2 Improvements in transport

Improvements in the transport of bulk cargoes (e.g. iron ore) and containers have lowered shipping costs, stimulating international trade and globalisation. Larger vessels have allowed carriers to achieve **economies of scale** and lower costs. Container ships can now carry up to 8,700 containers and the largest bulk carrying vessels weigh in at 350,000 tonnes. Thanks to these developments, it costs just US$10 to transport a television set from China to the UK and US$25 to ship a tonne of iron ore from Brazil to China.

2.3 Improvements in communications

Advances in information technology and telecommunications have stimulated a huge growth of international trade in services. Many TNCs in developed countries have transferred functions such as accounting, customer care and billing overseas — a process known as **international outsourcing** or **offshoring**. India has been particularly successful in capturing this trade, which is now worth US$25 billion a year to the Indian economy. In addition, highly skilled global ICT services have been offshored to India, making the southern city of Bangalore a world leader in software design. India has also attracted several global media services and news agencies. None of this would have been possible without the development of the internet and modern telecommunications.

2.4 Labour costs and skills

Many developing countries have attracted inward investment from TNCs because of their low labour costs and, in some cases, their high skill levels. India, for example, has 2 million English-speaking graduates who enter the labour market every year. A further attraction is the relatively low wages, on average just one-quarter of those of ICT workers in Europe and the USA.

The US–Mexico border region has experienced massive economic growth since the 1960s (Figure 12.2). This is due largely to investment by US, European and Asian manufacturing TNCs on the Mexican side of the border. Most investment comprises branch plants or *maquiladoras*, which manufacture a wide range of goods including clothes, chemicals and electronics. The main attractions of the border region are its low labour costs and proximity to US markets.

Labour-intensive industries such as textiles and clothing are particularly sensitive to labour costs. Typical wage rates for workers in the clothing industry in the Yangtze delta region of eastern China are US$1.4 per hour, compared to US$11.20 per hour in the UK.

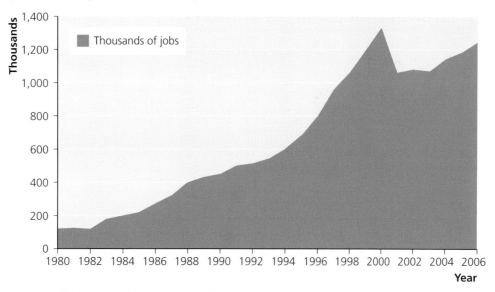

Figure 12.2 Maquiladora employment in Mexico, 1980–2006

2.5 Proximity to markets

Many large TNCs have globalised production by locating plants offshore. Since the 1980s, Japanese and Korean motor vehicles and electronics companies have invested in production facilities in North America and Europe (Figure 12.3). The advantages to foreign TNCs are:

- proximity to overseas markets
- avoiding trade barriers such as import tariffs and quotas
- the improved ability to respond to local market preferences in terms of design
- financial incentives offered by governments eager to attract inward investment and jobs

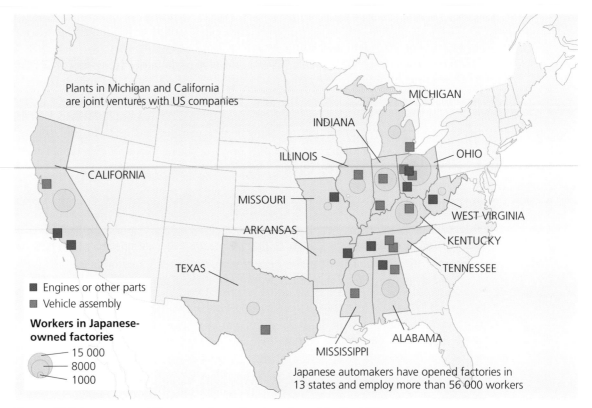

Plants in Michigan and California are joint ventures with US companies

MICHIGAN

INDIANA

ILLINOIS

OHIO

CALIFORNIA

MISSOURI

WEST VIRGINIA

ARKANSAS

KENTUCKY

TEXAS

TENNESSEE

■ Engines or other parts
■ Vehicle assembly

Workers in Japanese-owned factories

— 15 000
— 8000
— 1000

ALABAMA

MISSISSIPPI

Japanese automakers have opened factories in 13 states and employ more than 56 000 workers

Figure 12.3 Japanese and Korean automotive investment in the USA

3 *Transnational corporations (TNCs)*

3.1 Organisation and importance

Transnational corporations (TNCs) have been the dominant force in the globalisation of industry, services and trade. TNCs are large companies that either make or source products worldwide for sale in global markets. These companies often have complex structures, with regional organisation geared to local markets (Table 12.1).

Table 12.1 Organisation of IKEA and Toyota

	IKEA	Toyota
Headquarters	Leiden, close to Amsterdam. Amsterdam is a major urban centre with excellent communication and transport links.	Toyota City, Japan. There is easy access to Tokyo with its financial market and excellent communication and transport links. There are regional headquarters in all the main markets, with power to make local decisions.
Research, development (R&D) and design	Älmhult in southern Sweden is the main design centre. Design is also outsourced, mainly to Sweden.	R&D and design are concentrated within the company and are strongest in Japan. There are seven R&D units outside Japan, which are located in the main regional markets.
Production	Most production is outsourced. Subcontractors are located worldwide in 50 countries. IKEA has its own production arm — Swedwood — employing 13,000 people, mainly in Poland.	Production in Toyota's own assembly plants and parts factories is located in Japan, Asia, Europe, North America, South America, Africa and Oceania. Toyota buys in a large proportion of automotive parts from international subcontractors.

Car-making giants such as Ford and General Motors have adopted this type of structure, with North American, European and other divisions. Within these regional divisions, the companies source most of their parts from independent suppliers and manufacture models that reflect regional tastes and

preferences (Figure 12.4). Ford and General Motors also produce some models for the global market. This allows them to achieve economies of scale, producing fewer models but in greater numbers.

The dominance of TNCs in the global economy is summarised as follows:

- TNCs are responsible for four-fifths of global economic output
- the top 500 TNCs account for 90% of foreign direct investment (FDI)
- TNCs account for two-thirds of world trade: one-third of world trade is intra-firm trade between TNCs

3.2 Flexibility

TNCs take advantage of geographical differences in cost of the **factors of production** (e.g. natural resources, labour, capital) and of differences in state policies. Most countries encourage **foreign direct investment** (FDI) and, in competition with other states, offer incentives such as reduced taxes and subsidies to attract TNCs. However, the

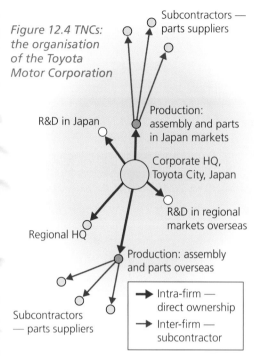

Figure 12.4 TNCs: the organisation of the Toyota Motor Corporation

relationship between states and major TNCs is not always equal. Given their flexibility, TNCs can switch production elsewhere where costs are lower or the financial inducements are more attractive. This can be done on an international and even on a global scale. For example, in 2006 Peugeot, the French TNC, moved production of the new 207 model to the Czech Republic, where labour costs were only one-fifth of those in the UK. The result was the closure of Peugeot's Coventry factory and the loss of 2,300 jobs.

4 TNCs and social and economic issues

4.1 Social and economic issues in developed countries

TNCs have the power to shape the global economy. They control scarce resources, which they can shift between states in terms of production. Because of their control of flows of capital, materials, components and technical expertise — all of which are crucial to economic development — individual governments are not inclined to challenge TNCs. However, although FDI has economic advantages, it also creates disadvantages (Table 12.2), and this can result in conflict and a range of social and economic issues.

Table 12.2 Advantages and disadvantages of TNCs and FDI to national economies

Advantages	Disadvantages
Provide inward investment and create jobs for local people.	Exploitation of workforce, especially in developing countries, with poor working conditions, low wages etc.
Increase incomes and raise living standards among employees.	Environmental pollution (in developing countries), which governments may tolerate to secure investment.
Boost exports and help the trade balance.	Lack of security, as TNCs switch operations to lower-cost locations elsewhere.
Develop and improve skill levels and expertise among the workforce and technology and process systems among local firms.	Lack of control, with key investment decisions taken overseas at company headquarters.
Increase spending and create a multiplier effect within local economies.	TNCs may demand further government financial incentives not to disinvest.

Attract related investment by suppliers and create clusters of economic activity.	In many developing countries, jobs are mainly low-skilled in labour-intensive industries (electronics, clothing).
	In developed countries, capital-intensive, foreign enterprises may offer few higher paid managerial, development, design and marketing opportunities.
	Competition can lead to the closure of domestic firms.

CASE STUDY 1   Hoover-Candy's white goods plant at Merthyr Tydfil, 2009

Background	Hoover-Candy is an Italian TNC with headquarters near Milan. The company employs 7,610 people worldwide, 6,000 outside Italy. It has 38 branch plants. It manufactured washing machines and tumble dryers at its Merthyr Tydfil plant. The plant was originally opened by the US TNC, Hoover, in 1948. Merthyr Tydfil was an early example of FDI and at its peak in the 1970s it employed over 5,000 workers.
Closure	The company closed its Merthyr Tydfil plant in March 2009. The reason for closure was its relatively high costs, making it uncompetitive with plants in China and eastern Europe. Production was transferred to Turkey, where costs are lower.
Impact	450 people worked at the plant. Although some operations remained on the site (e.g. warehousing, after sales service), closure of the plant resulted in 337 job losses. Hoover-Candy's demise marked the end of large-scale manufacturing in the town and Merthyr Tydfil was poorly placed to absorb the impact of closure. The largest employers are now the council, the health service, a T-Mobile call centre, the Welsh Assembly and Tesco. Unemployment has risen rapidly — up to 6.5% by the end of 2008 — with 16.5% of the workforce on incapacity benefit.
	The lack of employment exacerbates other problems. The town's population has fallen to 55,000 and it is the only district of Wales whose population is expected to shrink still further. By almost any measure — social well-being, health, educational attainment, wages, life expectancy, alcohol abuse, house prices — the town is close to the bottom of national league tables.

4.2 Social and economic issues in developing countries

Despite their economic success, TNCs have attracted widespread criticism, not least from anti-globalisation and environmentalist groups. This is because their business interests, especially in developing countries, often conflict with social, environmental and health concerns.

In the developing world, a major issue concerning TNCs and globalisation is the low pay and appalling **sweatshop** conditions of millions of factory workers. In China, hourly wages of factory workers are only a fraction of those in developed countries. Without overtime it is not unusual for factory employees to receive the minimum monthly wage of US$27 for a 40-hour week. Moreover, workers often sleep in dormitories attached to factories, while health and safety legislation within the factories is rarely enforced.

In lower-cost locations such as Vietnam, Cambodia and Bangladesh, pay and conditions are even worse. Already, as labour costs begin to rise in China, TNCs are **outsourcing** production to these countries.

In the past most TNCs have exerted little pressure on their suppliers to improve working conditions and pay, knowing that to do so would increase costs and reduce their competitiveness. Fila, a South Korean TNC and leading brand in sport shoes and other sportswear, is one such example. Like other leading sportswear companies, Fila does no manufacturing of its own. Instead, it subcontracts production to its suppliers in Asia, Latin America and eastern Europe. Because sportswear manufacture is labour intensive, Fila outsources production to countries with low-wage economies. Meanwhile, sweatshop conditions are created by TNCs pushing their suppliers for shorter delivery times and lower unit costs. Unless these targets are achieved, contracts may be cancelled and new suppliers recruited, threatening the existing workforce with redundancy.

CASE STUDY 2   Tae Hwa sports shoe factory, Indonesia

Background	The factory, located just west of Jakarta, makes sports shoes for Fila and other brands. It is owned by a South Korean company and employs 5,250 workers, 80% of them women.
Wages	The basic monthly wage at the legal minimum is $U72/month or 42 cents per hour. Workers also receive a small food allowance. Basic wages are too low to meet living costs.
Overtime	Workers can only earn a living wage by working overtime. Overtime is compulsory and illegally high. During periods of peak demand, employees work 13–14-hour shifts. Occasionally, they work 32-hour shifts and sometimes for a whole month without a day off. Workers refusing to do overtime can be demoted or sacked.
Output targets	The management sets impossibly high targets for the number of shoes to be completed in a day. To meet the targets, women work for an extra 1 or 2 hours a day unpaid. Unions are weak and workers may be blacklisted for union involvement.
Health and safety	Workers complain of needle-stick injuries and backache caused by repetitive strain. Many workers get sick from the chemicals used in the factory, which cause respiratory problems and allergies.

5   TNCs, globalisation and the environment

It is claimed that one reason why TNCs offshore manufacturing to developing countries is to exploit weak pollution laws and their lax enforcement. As a result, costs are passed to the environment as air and water pollution, indirectly exposing local people to ill health and disease.

Since the 1960s the US–Mexico border region has proved hugely attractive to FDI from US, European and Asian TNCs. A wide range of goods is manufactured on the Mexican side of the border including clothes, chemicals and electronics. Location in the border region gives the dual advantage of low labour costs and proximity to US markets.

However, large concentrations of industry in this area create environmental problems. Air pollution from hundreds of factories (and especially in the 14 twin cities that straddle the border) is responsible for respiratory disease, cardiovascular disease and premature death. The worst pollution is in the twin cities of El Paso (US) and Ciudad Juarez (Mexico). Ciudad Juarez has over 300 foreign-owned factories. Air pollution from particulates, ozone, carbon monoxide and nitrogen oxide exceeds Mexico's air quality standards and pollution drifts across the border and affects El Paso.

However, only a relatively small proportion of pollution comes from factories connected to foreign TNCs. Globalisation and the pace of economic growth is more to blame. Much of the pollution is from the city's 350 small, family-owned brick kilns, and dust from unmade roads and footpaths. Meanwhile, air quality standards in Mexico are roughly the same as in the USA, though enforcement is less rigorous.

CASE STUDY 3   The Pearl River Delta, China

Background	Guangdong in southeast China has been at the centre of China's recent drive to industrialisation and economic growth (Figure 12.5). The province has a population of 95 million and covers an area similar in size to England and Wales. At the industrial heart of Guangdong is the Pearl River Delta (PRD) Special Economic Zone, which is geared to the export of manufactured goods to North America and western Europe.
Industrial development	Globalisation has been the driver for industrial development in the PRD and Guangdong. The Chinese government has encouraged development by making the region a **free-trade zone**. This, together with its coastal location, has attracted major inward investment. Hong Kong is the main investor, followed by Taiwan, Japan, South Korea, Singapore and the USA. Many of the world's leading TNCs, including Hitachi, Samsung, Sony, Siemens, Pepsi and Toyota, have located production in the PRD. This has helped to make the province one of the most prosperous in China. In 2006 the PRD's GDP per capita was 2.5 times greater than the provincial average. Around four-fifths of Guangdong's GDP derives from the PRD.

| Environmental problems | Industrial development has stimulated massive in-migration (from the countryside) and urbanisation. In the PRD, 77% of the population lives in towns and cities. However, the provision of sewage treatment and industrial effluent has not kept pace with urban and economic growth. Three-quarters of Guangdong's cities have no sewage treatment plants, polluting rivers and threatening water supplies. The Pearl River is so grossly polluted downstream from Guangzhou that it is virtually lifeless. |
| | Air pollution is also a major concern. Sulphur dioxide from coal-fired power stations and manufacturing industries is above safe levels in Foshan and Guangzhou for several months of the year, especially in winter. Pollution from nitrogen oxide from motor vehicles, ozone and particulates shows a similar pattern (Figure 12.6). Guangzhou is affected by brown smog from motor vehicles 130 days a year on average, while acid rain, caused by air pollution, is widespread. |

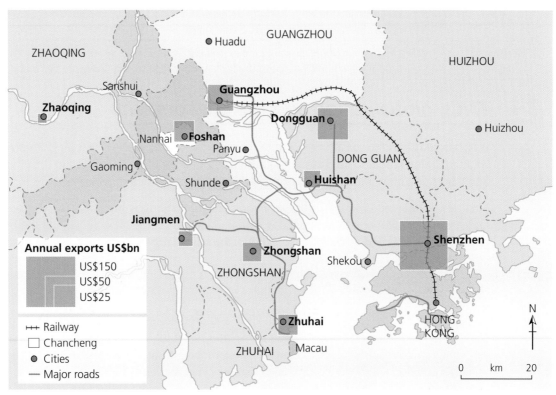

Figure 12.5 Guangdong, China

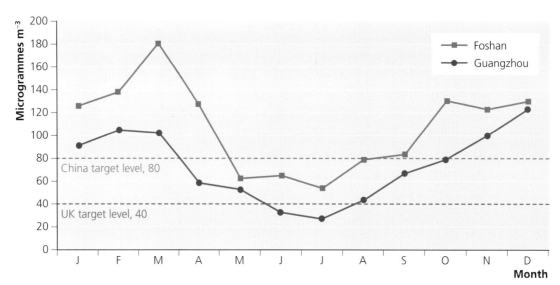

Figure 12.6 Pollution in the PRD

1 The development gap

1.1 The scale of the development gap

The **development gap** describes the difference in wealth between rich and poor countries. The gap is huge and has widened over the past 40 years (Figure 13.1). In 2007 average GDP per capita in the world's richest countries was over US$38,000 and life expectancy averaged 80 years. In the world's least developed countries (as defined by the UN), comparable figures were US$440 and 56 years. This means that on average people living in the richest countries were, in monetary terms, nearly 90 times better off and could expect to live roughly 50% longer than people in the poorest countries.

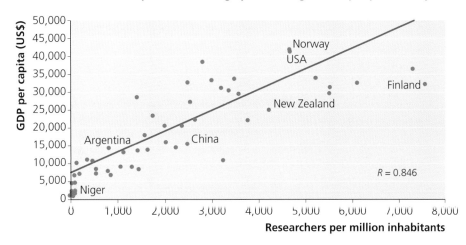

Figure 13.1 The development gap

The difference in wealth between the richest and poorest nations can be measured in many other ways, but all show the same wide development gap. There are two exceptions to this trend: China and east Asia. There, the development gap is narrowing. In 1970 east Asia's GDP per capita was just one-fifteenth that of North America. However, between 1970 and 2006, thanks to industrialisation and globalisation, east Asia experienced a 16-fold increase in its GDP per capita. As a result, by 2006 its GDP per capita had been narrowed to one-eighth that of North America.

1.2 The terminology of development

We use a number of different terms to distinguish rich and poor countries. The term **Third World** is still widely used when referring to poor countries. Originally devised during the Cold War of the 1950s to differentiate between capitalist (**First World**), Soviet (**Second World**) and non-aligned countries (Third World), the collapse of communism in the former Soviet Union made the idea of a Third World somewhat meaningless. At a more general level, the terms **rich North** and **poor South** provide an even simpler shorthand for global development. The so-called **north–south divide** is delimited by the **Brandt Line**, named after Germany's chancellor in the 1980s. However, in this dichotomy, the north–south division is not strictly geographical. Many of the world's poorest countries in Africa are in the northern hemisphere, whereas rich countries such as Australia and New Zealand are in the south.

Today, the most commonly used terms to describe rich and poor countries are **more economically developed countries** or just **developed countries**, and **less economically developed countries** or just **developing countries**.

However, the idea that we can divide all countries into two categories is simplistic. Development is a continuum, with rich and poor countries at the extremes of the distribution. Countries occupying the middle ground and currently undergoing rapid economic development are often called **newly industrialising countries** (NICs) or **emerging economies**. Examples include China, Brazil, India and Malaysia. Perhaps the clearest and most effective classification is based on income. Using this criterion, the United Nations recognises four categories of development: high income, upper middle income, lower middle income and low income countries (Table 13.1).

Table 13.1 Development indicators, 2007

Income	Life expectancy at birth (years)	Total fertility rate	Primary school completion (%)	FDI per capita (US$)	Malnutrition (% under 5s)
High income	79	1.8	99	1,527	0
Upper middle income	71	2.0	99	254	0
Lower middle income	69	2.3	91	70	25
Low income	57	4.2	65	25	28

Source: World Bank

2 Measuring development

2.1 GDP per capita

GDP per capita in US$ is the most widely used measure of economic development (Figure 13.2). It is calculated by dividing the total value of goods and services produced in a country (or region) in a year by its total population. To take account of international differences in the cost of living, it is adjusted for **purchasing power parity** (PPP).

As a measure of development, GDP does not include social factors such as the quality of education or healthcare in a country. It therefore tends to understate human well-being and the quality of life in socialist countries such as China and Cuba. Moreover, GDP per capita tells us nothing about the distribution of wealth and economic inequality within countries.

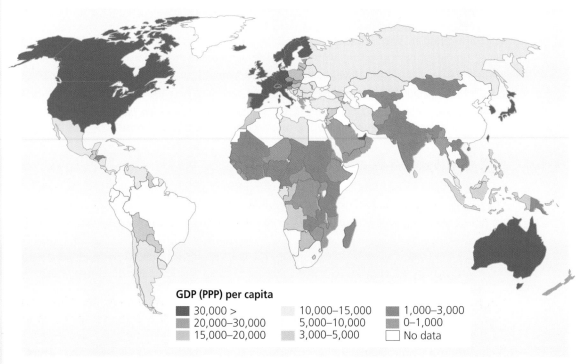

GDP (PPP) per capita

- 30,000 >
- 20,000–30,000
- 15,000–20,000
- 10,000–15,000
- 5,000–10,000
- 3,000–5,000
- 1,000–3,000
- 0–1,000
- No data

Figure 13.2 GDP per capita, 2008

2.2 Alternative measures of development

Alternative measures of economic development include **energy consumption per capita** (Table 13.2), **internet use** and **foreign direct investment per capita**. Infant mortality and life expectancy at birth are sensitive indicators of economic and social development. Both measures are influenced by economic factors such as income and poverty, as well as social factors like diet, sanitation and healthcare. The United Nations **Human Development Index** (Box 1), which uses a range of economic and social indicators, is increasingly preferred as the most accurate and comprehensive development measure.

Table 13.2 GDP per capita and energy consumption

Country	GDP per capita (US$)	Energy consumption per capita (tonnes of oil equivalent)
USA	44,155	7.56
Germany	35,270	3.78
Russia	6,632	4.87
Brazil	5,659	1.16
Argentina	5,471	1.85
China	2,034	1.51
Egypt	1,426	0.96
Philippines	1,382	0.26
Pakistan	810	0.39
Bangladesh	429	0.14

Box 1   *The Human Development Index*

The UN's Human Development Index (HDI) measures development using social and economic criteria. UN member states are ranked each year from 1st (most developed) to 179th (least developed). In 2008, Iceland was ranked 1st and Sierra Leone 179th. On the basis of the HDI index, 75 countries in 2008 were classed as having 'high' human development, 78 as 'medium' human development and 26 as 'low' human development. Among the countries with 'low' human development, all except one were in sub-Saharan Africa.

The HDI is a composite measure, combining data on life expectancy, adult literacy, school enrolment and GDP per capita (PPP in US$). It reveals huge inequalities in human well-being (i.e. the development gap) between countries. Also, because it includes both economic and social factors, it gives a more complete picture of development than indicators such as GDP per capita and life expectancy.

3   *Explaining the development gap*

There is no single explanation for international differences in levels of development and global inequality. The development gap is due to a combination of physical, historical, political, social and economic influences.

3.1 Physical influences

Natural resources such as energy, minerals and forests improve the prospects for economic development. Countries such as the USA, Canada, Russia and Saudi Arabia exemplify the advantages of a large natural resource base. However, abundant natural resources are not a pre-requisite for economic development. Some of the richest countries (e.g. Japan, Denmark, the Netherlands) have limited natural resources. Their success is based on human capital, especially the education and skills of their workforces.

Access to the oceans for trade (and, before the twentieth century, the diffusion of new technologies) has played a pivotal role in economic development. In Africa there are 15 **landlocked** countries with no direct access to the oceans and international trade. This disadvantage has undoubtedly contributed to their lack of development (Figure 13.3)

Climate and soils are important natural resources that influence food production and tourism. Many developed countries have humid, equable climates that favour farming. In contrast, many developing countries in the tropics and sub-tropics have arid and semi-arid climates that hold back development (e.g. Niger, Somalia). Development has also been hindered in the tropics by the prevalence of diseases such as malaria and yellow fever, which thrive in hot, humid conditions.

3.2 Historical influences

Most developed countries have a long history of economic development stretching back to the nineteenth century. Countries like the UK and France have over centuries exploited their own natural resources and built up human capital and infrastructure. This is in stark contrast to many countries in Latin America, Africa and Asia, where economic development only started in the twentieth century.

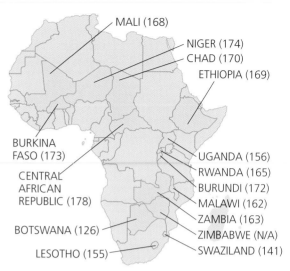

Figure 13.3 Africa's landlocked nations and 2008 HDI ranks

Development in many developing countries was held back by **colonialism** and its legacy. Nineteenth-century colonialism was exploitative. Colonies supplied food products and minerals for the benefit of the European imperial powers such as Britain, France and Spain. Economic development was only encouraged where it assisted resource exploitation (e.g. building ports and railways). Meanwhile, many modern developing countries in Africa were the artificial creation of colonial powers. Their borders (often following lines of latitude and longitude) paid little heed to internal tribal, national and cultural differences. The result — political instability and civil war — has put a brake on development in countries like Kenya, Rwanda, Nigeria and the Sudan.

3.3 Political influences

Poor governance has hindered dozens of developing countries in their drive to development, especially in Africa. Corruption is often widespread: money that should be used for development is often channelled to the military and governing elite. The recent history of Zimbabwe underlines the importance of good governance. Once the 'bread basket' of southern Africa, food production declined by one-third between 2000 and 2005, leaving nearly half the population malnourished and reliant on food imports and food aid. The cause was entirely political. Government land seizures destroyed the country's highly efficient farming industry, hitting food exports, food security and rural incomes. The extent of Zimbabwe's economic crisis is evident in its declining GDP per capita. It slumped from US$1,900 in 2004 to just US$150 in 2008, the lowest in the world.

The positive role that government can play in promoting development is illustrated in the modernisation and industrialisation of Japan in the late nineteenth century, and more recently in China. Economic change began in China in 1978. The government, led by Deng Xiaoping, aimed to raise standards of living by gradually moving away from a command economy to a more market-based one.

3.4 Social influences

Social influences on development include human capital, gender inequality and population growth. Although all three act causally to hold back development, we must recognise that each is also the outcome of low levels of development.

Human capital

Human capital includes the education and skills of the workforce and is a major driver of economic development. The world's most advanced economies rely on human capital as their principal resource. Pre-industrial economies in Africa and Asia create little demand for a highly educated workforce. With limited productivity and little engagement in international trade, GDP per capita remains low in these economies.

Explaining the development gap

Gender inequality

In many developing countries, religious and cultural practices and values effectively discriminate against women and neutralise a large part of the potential workforce. As a result, productivity is reduced and economic development is held back. **Gender inequality** is most pronounced in sub-Saharan Africa, south and southwest Asia and Arab states. Sub-Saharan Africa's poorest countries have the highest gender inequality, although the most extreme inequality is found in two Islamic states: Afghanistan and Yemen.

Rapid population growth

Many of the world's poorest countries have rapidly expanding populations with annual growth rates exceeding 2 % (Figure 13.4). In extreme cases, population growth exceeds economic growth, contributing to poverty and high rates of dependency. Even when economic growth outstrips **rapid population growth**, economic progress is inevitably slowed.

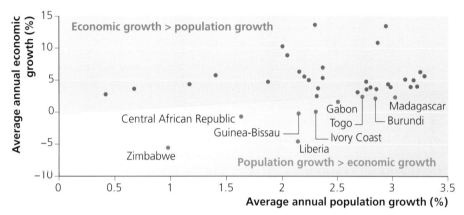

Figure 13.4 Expanding populations

3.5 Economic influences

The main economic factors that influence the development gap are **foreign direct investment** (FDI), **international trade** and **infrastructure**. FDI is crucial to economic development. It generates jobs that input spending and taxes to national economies, stimulates economic growth and attracts new skills and technologies (Figure 13.5). There are large global inequalities in FDI. They correlate closely with levels of economic development. For example the EU, with 7.6 % of the world's population, attracts 46 % of all FDI. On the other hand Africa, with 14 % of the world's population, receives only 3.4 % of FDI.

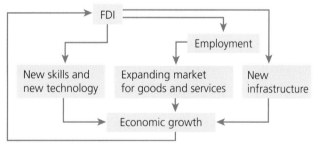

Figure 13.5 The positive economic impact of FDI

There is also a close correlation between international trade and economic development (Figure 13.6). International trade is dominated by developed countries and contributes significantly to the prosperity of countries like the USA, Germany and Japan. In contrast, Africa and Asia, with 75 % of the global population, control only 30 % of world trade in merchandise and services.

A pre-requisite for successful economic development is a modern infrastructure of roads, railways, international airports, deepwater harbours and telecommunications networks. An efficient infrastructure is essential to attract FDI, tourism and many other economic activities.

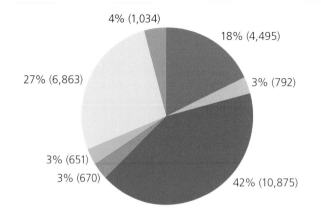

- North America
- Latin America
- Europe
- Russia
- Africa
- Asia
- Others

Figure 13.6 Value of international trade and economic development

4 Dependency theory

Dependency theory attempts to explain spatial differences in inequality and development. It suggests that that resources (i.e. materials, capital, people) flow from poor peripheral regions to wealthy core regions. Therefore, the core regions grow at the expense of the periphery while the periphery remains dependent on the core. Dependency theory can be applied at the scale of a single country or at the global scale. At the global scale, the core comprises the developed countries in North America, Europe and Japan. The periphery covers developing countries in Latin America, Asia and Africa. **Core-periphery inequality** develops from the working of free trade under a capitalist system. It is essentially a Marxist idea. Free marketeers on the other hand argue that free trade increases global wealth, which ultimately benefits us all. The cornerstone of dependency theory is Friedmann's core-periphery model (Box 2).

Box 2 *Friedmann's core-periphery model*

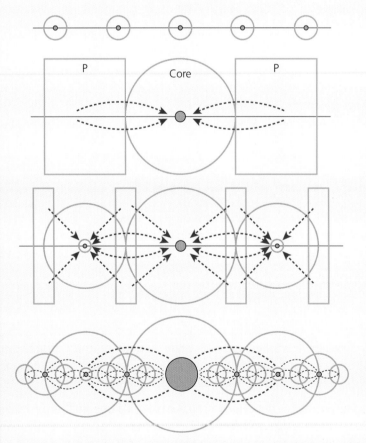

Stage 1 Relatively independent local centres; no hierarchy. Typical pre-industrial structure; each city lies at the centre of a small regional enclave.

Stage 2 A single strong core. Typical of period of incipient industrialisation; a periphery emerges; potential entrepreneurs and labour move to the core; national economy is virtually reduced to a single metropolitan region.

Stage 3 A single national core, strong peripheral sub-cores. During the period of industrial maturity, secondary cores form, thereby reducing the periphery on a national scale to smaller intermetropolitan peripheries.

Stage 4 A functional interdependent system of cities. Organised complexity characterised by national integration, efficiency in location, maximum growth potential.

Figure 13.7 Core-periphery model

Friedmann's **core-periphery model of economic development** describes four stages of spatial economic development (Figure 13.7). In the pre-industrial stage (stage 1), the geography of economic activity has a fairly uniform distribution. Early industrialisation (stage 2) creates regional inequalities with the emergence of a core and a periphery. The periphery exists in a state of dependency and its development lags behind the core. Resources and people flow from the periphery to the core, stimulating growth in the core and draining the periphery. As the economy expands in stage 3, wealth in the core slowly begins to spread to the periphery. This is due to rising costs of land and labour in the core and increasing levels of congestion. Finally in stage 4 spatial economic inequalities between core and periphery began to disappear, and wealth and economic activity are once again spread evenly.

Friedmann's model provides a conceptual framework for explaining the development gap. The present-day relationship between the economic core (developed countries) and peripheral regions such as sub-Saharan Africa and southwest Asia has parallels with stage 2 of the model. Many African countries rely heavily on the core, exporting minerals, energy, agricultural products and labour. These flows appear to benefit the core regions but hinder development in the periphery. The model also provides an explanation of emerging economies such as China and India and globalisation. High costs and skills shortages in many developed countries have stimulated offshoring and flows of FDI to the periphery, which are helping to close the development gap.

5 *Reducing the development gap*

Governments in developed countries, international agencies such as the United Nations and World Bank, and hundreds of **non-governmental organisations** (NGOs) such as Save the Children, Oxfam and WaterAid are committed to tackling global poverty and reducing the development gap. A range of strategies are used. They focus on the UN's **Millennium Development Goals** (MDGs), **development aid**, **international trade** and **debt relief**.

5.1 Millennium Development Goals

In 2000 the United Nations published its eight Millennium Development Goals (MDGs). They are to:
- eradicate extreme poverty and hunger
- achieve universal primary education
- promote gender equality and empower women
- reduce child mortality
- improve maternal health
- combat HIV/AIDS, malaria and other diseases
- ensure environmental sustainability
- develop a global partnership for development

The MDGs form a blueprint, agreed by all the world's countries and leading development agencies, to meet the needs of the world's poorest people. There are targets for each MDG, to be achieved by 2015. The targets include reducing by half the number of people living in absolute poverty (i.e. on less than a dollar a day) and reducing under-5 mortality by two-thirds (both compared to 1990 levels).

5.2 Development aid

Development aid describes capital, technical assistance and other resources transferred from developed countries to developing countries. The donor countries or organisations do not expect full or direct repayment. This type of aid is long term.

Donors are mainly rich countries that belong to the Organisation for Economic Co-operation and Development (OECD). The notional target for development aid, set by the UN, is 0.7% of GDP. However, few countries outside Scandinavia achieve even this modest target. In 2008, total development aid from the 22 OECD countries was nearly US$120 billion. This represents 0.3% of GDP for that year.

Box 3 *The World Bank and types of development aid*

The World Bank was set up in 1944 at the same time as the **International Monetary Fund** (IMF). Initially, the World Bank was charged with financing the construction and development of major infrastructures in developing countries such as dams and roads. In total, the World Bank has provided US$50 billion for large dam construction in 92 countries. This **'top down'** approach has not always been of direct benefit to the poorest people and has prompted a rethink. Now the World Bank increasingly targets smaller schemes linked to raising incomes, increasing food production and improving water supply, education and healthcare at a local level.

There are two types of development aid: bilateral aid and multilateral aid. **Bilateral aid** describes the assistance of a donor government to a **recipient** country. In 2007, nearly 70% of all foreign aid was bilateral. **Multilateral aid** operates through donor agencies like the World Bank, International Monetary Fund and the United Nations, as well as non-governmental organisations. These organisations are largely funded by the rich OECD countries.

CASE STUDY 1	The Nam Theun 2 dam, Laos
Background	The scheme involves large-scale development aid in the region of US$1.3 billion to construct a dam on the Nam Theun River (a tributary of the Mekong) in Laos, southeast Asia, which is due to be completed in 2009. The scheme is funded by loans from the World Bank and the Asian Development Bank. It will generate 1000 MW of electricity, most for export to neighbouring Thailand.
Impact	The dam will saddle Laos with a substantial foreign debt. The reservoir created by the dam will displace 4,500 people, flood 450 km^2 of land and destroy habitats and wildlife in one of southeast Asia's largest remaining areas of rainforest (over 400 species of birds and 50 species threatened with extinction live in the forest). The livelihoods of 100,000 people who live downstream from the dam and who rely on the river for fish, drinking water and irrigation could be threatened. Fish migration could be disrupted and gardens currently irrigated for farming will disappear. Exported electricity (worth US$150 million a year) will provide income that, in theory, should help to alleviate poverty in Laos, one of the poorest countries in Asia. It should help to improve health, education and basic infrastructure.
Issues	Do large-scale schemes of this type benefit the poorest people whose lives are most disrupted? Do the economic benefits outweigh the social, economic and environmental costs?

In the 1970s and 1980s much development aid comprised large construction projects designed to improve infrastructure and energy supplies. Economists believed that the benefits of these schemes would eventually 'trickle down' and improve the lives of the poor. In reality, the benefits went largely to governments, urban areas and foreign companies.

Today, the emphasis has shifted to small-scale development aid projects funded by non-governmental organisations, government departments and multilateral development agencies. They are seen as a more effective way of directing aid to those most in need. This **'bottom up'** approach starts at the grass-roots level. It supports involvement by local people and uses their skills and labour, encouraging them to take ownership as stakeholders. Typical small-scale aid projects include improving irrigation and farming techniques, reafforestation programmes, sinking village wells to provide clean drinking water, improving sanitation systems, building schools and microfinance schemes.

Why do countries give development aid?

The motives of aid donors are economic, political and ethical. For recipient countries, development aid provides a number of economic advantages, such as:

- foreign investment
- capital to pay interest on foreign loans
- foreign exchange to pay for imports
- capital to fund the development of essential infrastructure
- additional income for government

5 *Reducing the development gap*

Development aid is also given for political reasons. Donor countries may be able to extend or maintain their political influence in an economically strategic region. China's development aid to the Sudan and the Democratic Republic of Congo gives it privileged access to these countries' rich natural resources.

In the past, much bilateral aid targeted at large-scale infrastructural projects was **tied**. Grants and loans came with conditions on how and where the aid should be spent. Invariably, this meant importing resources, machinery and technical assistance from the donor country. Tied aid provides economic advantages that benefit the donor country as much as the recipient. Countering this view, donor countries maintain that tied aid is needed to prevent:

- fraud by corrupt government officials
- the diversion of aid to military spending and extravagant projects such as presidential palaces and lavish parliamentary buildings

Humanitarian reasons also lie behind much development aid. Rich countries genuinely want to help people in developing countries escape poverty and achieve a better quality of life. Aid is also one way in which former colonial powers such as the UK and France can help repay some of the debt they owe to developing countries for past colonial exploitation.

5.3 International trade

Patterns of international trade

There has been a massive increase in the volume of world trade in the past 20 years. Between 1996 and 2006 the total value of merchandise and service exports more than doubled. The spectacular growth of world trade is associated with powerful growth in the global economy, globalisation and the removal of barriers to **free trade**.

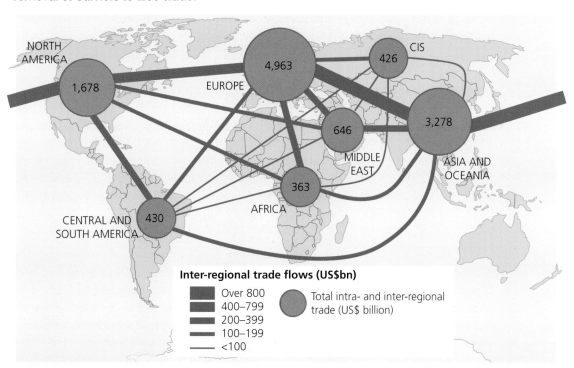

Figure 13.8 European trading partners

International trade has a number of features:

- Most international trade is intra-regional. For example, 74% of international trade in Europe is between European countries (Figure 13.8).
- Most inter-regional trade is between Europe, North America and Asia.
- International trade is dominated by developed countries in Europe, North America and Japan and by emerging economies such as China.

- The share of world trade of the poorest nations has declined since 1980.
- The rapid growth in international trade in services has been concentrated in developed countries.
- The top 500 TNCs account for 70% of total world trade.

International trade often disadvantages the world's poorest countries. There are several reasons for this:

- The trend towards free trade exposes the economies of vulnerable developing countries to competition. Agricultural exports such as bananas from the Caribbean and cotton from West Africa have been hit hard by producers that are either more efficient or that benefit from hidden subsidies from the governments in rich developed countries.
- Many developing countries rely on a narrow range of primary export commodities (e.g. mineral ores, coffee, tea, cotton). World prices for these commodities fluctuate more than prices for manufactured goods. This results in a dramatic loss of export earnings during periods of oversupply and low demand.
- World prices for manufactured goods have increased primary commodity prices, weakening the trade balance of many developing countries.

Box 4 *Regional trade blocs*

Many countries combine to form regional **trade blocs** such as the European Union (EU) and the North American Free Trade Agreement (NAFTA). One-third of the world's trade takes place within regional trade blocs. The most powerful trade blocs are in the developed world. Trade blocs promote the interests of member states by encouraging free trade between them and protecting their industries and services from foreign competition by using tariffs, quotas and subsidies. Within the EU there is free movement of goods, services, capital and people. There are therefore no tariffs, custom duties or taxes on the flow of goods within the EU. However, foreign goods entering the EU pay a common external tariff. Some EU industries such as agriculture and coal are heavily protected from foreign imports by price subsidies.

Fairer trade

Close political and historical ties between the UK (and other EU countries) and small banana-producing nations in Africa, the Caribbean and Pacific (ACP) has led to preferential trade agreements between the EU and these countries. Without preferential treatment, small producers would be unable to compete with larger, more efficient growers in Latin America whose operations are owned and conducted by major US TNCs (e.g. Dole, Chiquita). The current trade agreement between the EU and ACP consists of a 775,000 tonnes tariff-free banana quota. At the same time, the EU discriminates against rival Latin American producers by imposing a €175/tonne tariff on their banana exports. In 2007 the World Trade Organization (WTO) ruled that the agreement between the EU and ACP violated global trade rules, giving ACP growers an unfair advantage. This dispute between fair trade and free trade currently remains unresolved.

Developing countries argue that the policy of rich countries in providing farm subsidies reduces the value of their agricultural exports and violates the principle of fairness in trade. In 2008 the WTO declared the US government's US$3 billion subsidy to its cotton growers illegal. The result was a major victory for poor West African countries such as Burkina Faso and Senegal that have been harmed by US subsidies. For many farmers in these countries, cotton is their only source of cash. However, US cotton subsidies lower the world price of cotton, which means even less income for some of the poorest farmers in the world. Under the new WTO ruling, the world price of cotton should increase by 6 to 14% and West African farmers should get 5 to 12% more for their cotton. This will represent a significant boost to household incomes and help to lift thousands of families out of absolute poverty.

In the past 10 years fair trade products such as coffee (e.g. Cafédirect), tea, chocolate (e.g. Divine) and bananas have been successfully marketed in developed countries. Nearly 100 fair trade products are available in the UK. Instead of formal trade agreements, the fair trade idea gives consumers an opportunity to help poor farmers in developing countries. By paying a higher (i.e. fairer) price, farmers' co-operatives in developing countries receive extra fair trade money, which can be spent on individual families or community projects. Fair trade products have proved popular in developed countries and business is expanding rapidly. In the year 2007–08, for example, global sales of fair trade tea and coffee doubled. Fair trade products overall grew by 22% and were worth nearly US$3 billion.

5.4 Debt relief

Debt relief is one part of a much larger effort to tackle the development needs of low-income countries. In 1996 the IMF and World Bank launched the **Heavily Indebted Poor Countries Initiative** (HIPCI). It aims to ensure that no poor country faces a debt burden it cannot manage. However, for debt reduction to have a real impact on poverty, the additional money must be spent on programmes to benefit the poor. In 2005 the HIPCI was supplemented by the **Multilateral Debt Relief Initiative** (MDRI), which allows for 100% relief on debts owed to three multilateral institutions: the IMF, the World Bank and the African Development Fund. The **G8** heads of state agreed to assume full responsibility for these debts.

In order to qualify for debt relief, countries must meet certain criteria. They must commit to poverty reduction through policy changes and spend additional money on programmes that benefit the poor. They must also preserve peace and stability and improve governance and the delivery of basic services. By 2009, 24 countries had met these criteria and were receiving full debt relief. A further 16 countries are hoping to qualify.

1 Sources of energy

There are two major categories of energy resources: **non-renewables** (finite) and **renewables**. Finite or non-renewables comprise **fossil fuels** such as coal, oil and natural gas; renewables are more diverse and include wind power, hydroelectric power (HEP), solar power, biofuels and several others (Figure 14.1). About 81% of total world energy supply in 2007 came from coal, oil and gas; 10% consisted of combustible renewables (mainly wood) — the main source of energy for millions of rural dwellers in the developing world.

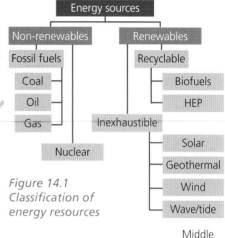

Figure 14.1
Classification of
energy resources

1.1 Fossil fuels

Oil is the leading fossil fuel. World production in 2007 was 3.937 billion tonnes. Its geography of production is highly concentrated, with 70% of global output originating in the Middle East, the OECD countries (mainly the USA) and the former USSR (Figure 14.2). Output is small in Australia and the Pacific region, Europe and south Asia.

Coal deposits are more widespread than oil. With the exception of South America, all the continents have substantial coal reserves (Table 14.1). In 2007 coal output exceeded 5.5 billion tonnes and provided just over one-quarter of the world's primary energy and 41% of the world's electricity (Figure 14.3). World coal production increased by 60% between 1986 and 2006, mainly because of the massive growth in output in the emerging economies of China and India.

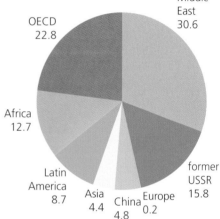

Figure 14.2 Global oil production,
2007 (%)

Table 14.1 Global coal production, 2007

Country	Coal produced (millions tonnes)
China	2,549
USA	981
India	452
Australia	323
South Africa	244
Russia	241
Indonesia	231
Poland	90
Kazakhstan	83
Colombia	72

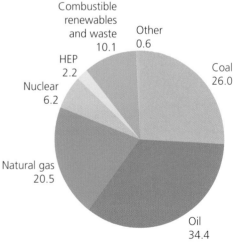

Figure 14.3 World energy production,
2007 (%)

Natural gas accounts for one-fifth of global primary energy production. At the global scale, production is widely dispersed (Figure 14.4). In the past 30 years production has increased more rapidly than either oil or coal. This is partly because:

- gas reserves are large (especially in Russia and the Middle East)
- gas is easier to transport and store (by pipeline and as liquid natural gas)
- gas is a cleaner and more environmentally acceptable fuel than coal and oil

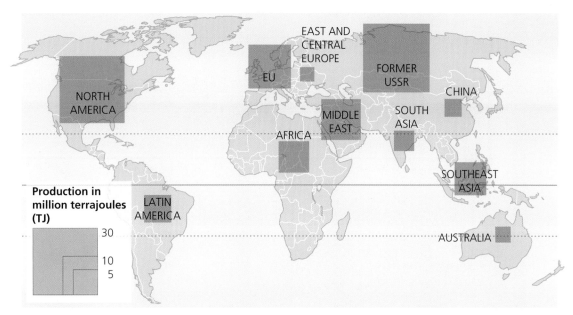

Figure 14.4 World natural gas production, 2007

1.2 Nuclear energy

Nuclear power produces 6% of the total world energy and 15% of the world's electricity. Because of the advanced technology needed to build nuclear plants and the need to control nuclear weapons' proliferation, most nuclear power plants are located in the developed world. Countries like France and Japan, which have limited domestic reserves of fossil fuels, have invested heavily in nuclear power. Together, France, Japan and the USA account for over half of global nuclear power production. Uranium fuels the nuclear energy industry. Like fossil fuels, uranium is a finite resource, although it can be recycled by reprocessing. In contrast with oil, uranium reserves are mainly found in politically stable countries such as Canada, Australia and Namibia. Reserves (3.3 million tonnes) are sufficient to last for several decades.

1.3 Renewable energy

Around 13% of the world's primary energy production is renewable, although most of this is wood used for cooking and heating in developing countries. Only 2.8% of global energy comes from alternative energy sources such as wind, solar, geothermal and HEP, of which four-fifths is derived from HEP. Therefore, although massive investment is currently going into alternative energy in developed countries (especially wind power), its overall contribution to global energy production will remain small for the foreseeable future.

2 Energy use and economic development

As wealth and economic development increase, a similar growth in energy consumption per capita occurs (Figure 14.5). Large amounts of energy are needed to sustain rising prosperity and high standards of living. For example, it takes on average 23 times more energy per capita to support a US consumer than one living in India. Energy is used to manufacture goods, construct buildings and provide transport, schools, heating etc. The reason why energy consumption is relatively low in developing countries such as India is poverty. People in the developing world depend mainly on goods and services produced locally. They cannot afford to buy expensive manufactured goods that require lots of energy to make and they cannot afford to own cars. Cooking and heating depend on biofuels like wood, and crops are often produced using human labour and draught animals rather than diesel-powered machinery. Poverty also means that most people have little if any disposable income. There is therefore little demand for international travel, which is extremely energy intensive.

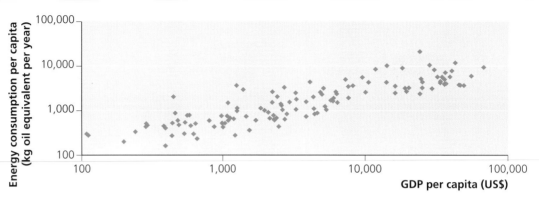

Figure 14.5 Energy consumption per capita

Box 1 *Energy mix*

Energy mix is the combination of energy types used to power a country's economy. Four main factors influence a country's energy mix: domestic energy resources, energy security, economic development and technology, and environment.

■ **Domestic resources**: exploiting domestic energy resources is often cheaper and more secure than relying on imports. The reliance of China and India on coal largely reflects the huge coal reserves found in both countries. Norway's use of HEP is explained by its mountain relief, high precipitation, snowfields and icefields.

■ **Energy security**: countries may opt to diversify their energy supplies to avoid excessive dependence on imports. This becomes important when imports like oil are sourced from politically unstable regions like the Middle East or where exporters could use energy as a political weapon (e.g. Russian gas).

■ **Economic development and technology**: in much of the developing world, poverty and low technology force people to rely on traditional sources of energy, especially biofuels such as timber, plant residues and animal dung.

■ **Environment**: many developed countries are investing in more sustainable forms of energy as they aim to cut carbon emissions. Environmental considerations have driven the recent growth in electricity derived from wind power in countries such as Denmark, Germany and the UK.

CASE STUDY 1	Sweden's energy mix
Economic development	Sweden is one of the world's most advanced economies. In 2006 its GDP was the 8th highest in the world. On the UN's Human Development Index (HDI) it ranked 5th highest. Swedes enjoy a high standard of living, which is supported by large quantities of energy consumption per person.
Energy mix	Sweden's energy mix comprises nuclear power (38%), oil (29%), HEP (26%), coal (5%) and natural gas (2%).
Factors controlling energy mix	This energy mix is closely related to Sweden's energy resource base. There are significant HEP resources on its major northern rivers (e.g. Luleälv, Umeälv), which in terms of their high discharge and natural lakes have enormous potential. A lack of indigenous fossil fuels explains Sweden's heavy commitment to nuclear power. Considerations of energy security in the 1960s also played a part in the development of the nuclear industry. Nuclear power would have made a larger contribution but for 30 years expansion was blocked by public opinion, opposed to nuclear energy on environmental grounds. Similarly, HEP expansion has been halted for environmental reasons: several wild northern rivers, not harnessed for HEP (e.g. Kalixälv, Tornälv), are protected from development by law.

3 *Energy security*

Energy security is about creating a situation where energy supplies, which are vital to the operation of a country's economy, are assured. A range of economic, political and environmental factors influence energy supply. When a country has energy security, it can:

- control these factors
- absorb changes in energy supply and availability with minimal damage to its economy

3.1 Global energy security

Economic influences

For decades there have been fears that the world will run out of crude oil. The latest estimates of global oil reserves (186 billion tonnes) suggest that, at current rates of consumption (4 billion tonnes a year), reserves will be depleted within 50 years. However, such forecasts are speculative because of:
- the discovery of new reserves
- changing rates of consumption

Oil producers have formed a powerful cartel called the Organization of the Petroleum Exporting Countries (OPEC), which can influence world oil prices by controlling production levels. Oil prices are also extremely volatile. In 2008 they reached a peak of US$131/barrel in July, only to fall to US$39/barrel by December. Global reserves of other fossil fuels, particularly coal, are larger and more secure.

Political influences

Because cheap energy is vital for economic activity, energy supplies can be used for political purposes. Over half of the world's oil reserves are in the Middle East — in Saudi Arabia, Iran, Iraq and the United Arab Emirates (UAE). This region has been plunged into political turmoil on many occasions in the past 60 years (e.g. the 1973 Israel–Arab war, the Iraq war), threatening global oil supplies. Moreover, some major Middle East producers are hostile towards the USA and the EU and in the past have deliberately cut off supplies.

Environmental influences

The combustion of fossil fuels, releasing carbon dioxide and other greenhouse gases (GHGs), is the main driver of global warming and climate change. Therefore, the detrimental environmental impact of coal, oil and natural gas puts in doubt the future sustainability of powering the global economy with fossil fuels. As a result, many developed countries are looking to alternative energy sources, especially renewables such as wind, solar and geothermal energy.

3.2 National energy security: the UK

In 2006 the UK produced nearly 197 million oil equivalent tonnes of energy from domestic resources. North Sea oil yielded 84 million tonnes, natural gas 80 million tonnes, coal 11.4 million tonnes and primary electricity (mainly nuclear) 17.7 million tonnes. Energy consumption in the same year was 232 million tonnes. This left an **'energy gap'** of 35 million tonnes, which was filled largely by imports of coal and gas.

The UK government is concerned to ensure security of energy supply in future and this consideration is an important influence on its energy policies. The key issues are:
- maintaining a diversified mix of energy sources, so avoiding excessive reliance on any one source
- filling the 'energy gap'. The gap is set to widen in the next few years for two reasons:
  - North Sea oil and gas production, now well past their peak, will decline further
  - most of the UK's ageing nuclear power stations are nearing the end of their productive lives (Figure 14.6).

Excessive reliance on Russian gas, delivered to western Europe through the trans-Siberian pipeline, would weaken the UK's energy security. In January 2009 Gazprom, the Russian gas supplier,

Figure 14.6 UK nuclear power stations

SCOTLAND
Hunterston   Torness
Chapelcross   Hartlepool
Heysham
Wylfa   ENGLAND
WALES   Sizewell
Bradwell
Oldbury
Hinkley Point   Dungeness

unilaterally shut off supplies to neighbouring Ukraine, which led to gas shortages in Germany and other EU countries. This action illustrated the danger of over-reliance on Russian energy supplies. It is this fear, together with the UK's international obligations to reduce carbon dioxide emissions, that has brought nuclear energy back into favour. Eleven potential sites have been identified, all of them on or near existing nuclear sites. The first new stations are expected to come into use in 2017.

3.3 The potential impact of energy insecurity

The potential impacts include:
- complex infrastructures that link producers and consumers
- developing new energy reserves
- the economic power and political influence of energy TNCs and OPEC countries

Energy infrastructure

The principal international flows of crude oil are shown in Figure 14.7. Elaborate energy infrastructures comprising oil and gas installations, shipping routes and pipelines transfer oil and gas from producer to consumer countries. In politically sensitive regions such as the Middle East and the Niger delta, these infrastructures are targets for terrorist groups.

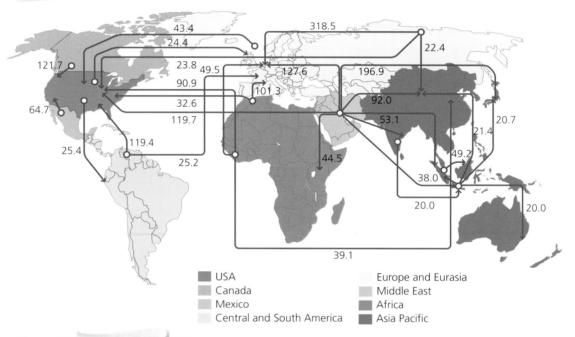

■ USA	Europe and Eurasia
Canada	Middle East
Mexico	■ Africa
Central and South America	■ Asia Pacific

Figure 14.7 Oil trade flows, 2008

Saudi Arabia's oil installations are an obvious target. Over half of the country's oil reserves are concentrated in just eight fields, including Ghawar, which provides half of the country's total oil production. Two-thirds of Saudi Arabia's crude oil is processed in a single enormous refinery at Abqaiq, 40 km inland from the Gulf of Bahrain. On the Persian Gulf, Saudi Arabia has just two primary oil export terminals: Ras Tanura and Ras al-Ju'aymah. Its Red Sea oil terminal at Yanbu is connected to Abqaiq by a 1,000 km pipeline. An attack on Abqaiq or Ras Tanura could reduce by half production of Saudi oil for at least 6 months, with disastrous economic consequences for major importers like the USA, the EU, Japan and China.

Increasingly, terrorists and pirates operate in busy shipping lanes, hijacking tankers and other vessels, especially at key 'chokepoints'. The most important are the Strait of Hormuz (which connects the Red Sea to the Gulf of Aden and the Arabian Sea, and through which 13 million barrels of oil pass daily) and the Strait of Malacca (between Indonesia and Malaysia). Nearly two-thirds of the world's oil is shipped through a small number of 'chokepoints' that are highly vulnerable to piracy and terrorism.

Meanwhile, about 40% of the world's oil flows through pipelines that are vulnerable to attack. Saudi Arabia, for example, has 15,000 km of pipeline and Iraq has 4,000 km. Most of the network is above ground and easily sabotaged. Because of their length, pipelines are difficult to protect and are therefore easy targets for terrorists.

CASE STUDY 2	Oil spillages in the Niger delta
Background	Nigeria is Africa's leading oil producer and 4th largest exporter. Reserves of oil (and gas) are concentrated in the Niger delta and 6,000 km of oil and gas pipelines cross the region. The delta supports Nigeria's largest surviving areas of rainforest and important mangrove forests. These ecosystems are extremely biodiverse and productive.
Environmental impact of oil spillages	There have been over 4,000 oil spills in the delta region since production started in 1958, with a loss of 500 million tonnes of oil. These spills have degraded the environment, destroying wildlife, ecosystems and the livelihoods of local people (pollution of water supplies, destruction of marine life). The situation is worst in Ogoniland in the southernmost part of the delta. This area has 100 oil wells, two refineries, a petrochemical complex and a fertiliser factory.
Causes	The pipeline network (some pipelines are 40–50 years old) is mainly above ground. It corrodes quickly in the hot, humid climate and poor maintenance is responsible for some spills. Spillages also occur through illegal tapping. Sabotage is also to blame for oil pollution. Local people have derived few benefits from the oil riches of the delta, while bearing most of the costs. This has spawned militant groups such as the Movement for the Emancipation of the Niger Delta (MEND), which has repeatedly targeted Shell and Chevron pipelines and oil pumping stations located in communities hostile to foreign oil companies. An attack in June 2009 caused the shutdown of Chevron's pipeline with a loss of 100,000 barrels a day.

Developing new energy reserves

The controversial decision by the US government to drill for oil and gas in the pristine Arctic National Wildlife Refuge (ANWR) in Alaska was taken by the Senate in 2005. Despite the impact on the fragile tundra ecosystem, the government argued that development was in the national interest. Specifically it would give greater energy security and reduce reliance on oil producers in the Middle East. However, the environmental costs are considerable. Drilling areas on the north slope are close to the main calving grounds of the caribou herds, while the breeding of millions of migrant birds in the high arctic during the summer could be disrupted. Moreover, the culture and subsistence way of life of indigenous people, who rely on hunting caribou and marine mammals, could be permanently harmed.

The main argument for opening up oil reserves in the ANWR is to enhance energy security by reducing the USA's reliance on OPEC and at the same time gaining greater control over oil prices.

In the UK the gradual depletion of North Sea oil and gas resources has increased energy imports and, as a consequence, **energy insecurity**. This concern intensified the search for new oil and gas fields in the North Atlantic, west of Shetland. The Clare oilfield, 75 km west of Shetland, was discovered in 1977, but production only began in 2008. Further west are the Foinaven and Schiehallion oil and gas fields operated by BP. Weather and sea conditions are severe, and the most advanced technology is needed to extract oil and gas in water 400–600 m deep. Environmental concerns in this 'Atlantic frontier' include the impact of noise from oil platforms and drilling rigs on cetaceans (whales, dolphins).

The bitumen tar sands in Alberta, Canada are a potentially huge source of oil, with estimated reserves equal to the entire oil reserves of Saudi Arabia. Large oil reserves are also found in oil shale deposits in Canada and the USA. The development of these resources is attractive, not least because it would restore North America's self-sufficiency in oil. The environmental impact caused by oil sand extraction is, however, considerable. It requires the removal of 16 tonnes of oil sand to produce 1 tonne of oil. Open-cast mining causes massive land damage, with the destruction of farmland, forests and entire ecosystems. By 2007 mining had disturbed 420 km^2 of land. Mining operations also give rise to concerns about air and water pollution.

Energy TNCs and cartels

Energy TNCs such as BP, and cartels of energy producers like OPEC, are major players in the global energy system. For this reason they exert a huge influence on economic policies and the global economy. Four of the ten largest TNCs are oil companies (BP, Exxon, Shell, Total). Their strategic and political importance is evident in their influence on governments. For example, the decision by the US government to allow drilling for oil in Alaska's Arctic National Wildlife Refuge was influenced by donations made by the oil industry to the Republican-controlled Senate and the president. Furthermore, the US government's denial of any link between global climate change and human activities was attributed to the power and influence of the oil and coal lobby in the USA.

OPEC is the cartel representing the major oil exporting countries. The 13 OPEC countries produce around 40% of the world's oil and hold around 75% of global oil reserves. By controlling production, OPEC regulates the amount of oil entering international trade and exerts an important influence on prices. Oil prices are a major driver of inflation and economic activity in the global economy.

4 Sustainable energy supplies

Although non-renewable energy sources such as coal, oil and natural gas cannot be managed sustainably, much can be done to reduce their consumption and improve their efficiency. Meanwhile, investment in sustainable energy sources such as HEP, wind and solar power can be expanded. International obligations to reduce carbon emissions under the **Kyoto Protocol** are a major impetus to move to a more sustainable use of energy. To achieve the carbon emissions targets set by Kyoto, countries will need to:
- reduce energy consumption
- rely on less polluting fossil fuels and develop alternative sources of energy

Under the Kyoto Protocol, the UK must achieve a 12.5% reduction of carbon emissions below 1990 levels by 2012.

4.1 Reducing energy consumption

The first step to achieving a sustainable energy economy is to minimise energy consumption. Governments can legislate to reduce emissions from motor vehicles, factories, commercial activities and households. In the UK, smaller cars with more efficient engines pay lower road tax and government grants are available for cavity wall and loft insulation, which help to conserve energy. Governments may also provide financial incentives to promote the use of non-petroleum fuels such as **biodiesel**, which is sustainable, and **liquid natural gas** (LNG), which is less damaging to the environment. In California a carbon tax is levied on electricity generated in coal-fired power stations to encourage the use of cleaner fuels such as gas or alternative energy like wind and solar power.

Carbon trading also helps to reduce energy consumption as well as favouring less polluting energy sources. Under this scheme, governments allocate credits to companies and organisations, allowing them to emit up to a specific amount of carbon dioxide a year. If the pollution quota is exceeded, the polluters must purchase credits from those companies that have remained within their quota. Therefore, carbon trading provides a financial incentive to conserve energy and pollute less. An international market in carbon trading, worth US$60 billion, has been established.

4.2 Alternative energy

The leading sources of renewable energy are HEP, biofuels, wind power, solar power and geothermal power. In 2006, 12.9% of the world's primary energy production came from renewables, of which 80% was biofuel, mainly wood. HEP contributed 2.2% and advanced technology alternative energies less than 1%.

Hydroelectric power

Hydroelectric power (HEP) is the most popular advanced technology type of alternative energy. In Europe it figures prominently in the energy budgets of Norway, Sweden, Switzerland and Austria.

Several developing countries have major HEP schemes. They include the Three Gorges scheme on the Yangzte in China, Itaipú on the Paraná in South America, and Aswan on the Nile in Africa, and Nam Theun in Laos, southeast Asia (see Topic 13). Ideal physical conditions for HEP include:

- powerful rivers with high annual discharge (large catchments, high precipitation)
- steep gradients (e.g. overdeepened valleys in glaciated uplands)
- water storage in natural lakes, glaciers and icefields

HEP has three main advantages. It is renewable, carbon-free and has the potential to generate enormous amounts of electricity (e.g. Three Gorges 18,200 MW, Itaipú 12,600 MW). However, there are issues concerning the sustainability of HEP. Most schemes involve the construction of dams, which can have adverse effects both upstream and downstream. Upstream vast areas are flooded, destroying habitats, wildlife, farmland and even settlements. Downstream the character and behaviour of the river are changed (e.g. colder clearer water, absence of floods), modifying habitats and damaging fluvial ecosystems. Additional problems include the huge capital costs of large dams and the questionable benefits derived by local people whose livelihoods and homes may disappear.

Wind power

Wind power is renewable and does not produce carbon emissions. The amount of energy generated depends on wind speed, air density, the spacing of wind turbines and the size of turbine blades. The UK, which is committed to generating 15% of all its electricity from renewables by 2020, will rely mainly on the expansion of wind power to meet this target. However, there is substantial hostility to the development of wind farms in the UK from conservationists and local people.

Wind farms require large tracts of land: 7 ha for every megawatt of electricity produced. They are visually intrusive, especially as they are often sited prominently on hillsides and ridges where wind speeds are high. Many of the best sites are on coasts and uplands, which have considerable environmental and amenity value. Wind farms also disturb wildlife habitats, and bird mortality caused by collisions with turbines and transmission lines is high.

Because of the difficulty in getting planning permission for wind farms on the mainland, the industry has responded by siting new wind farms offshore — in the Thames estuary, the Wash, Morecambe Bay and the Outer Hebrides. A new generation of giant wind turbines with blades up to 150 m in diameter will be sited several kilometres offshore, where they are visually less obtrusive.

Solar power

Solar-powered electricity can be produced in two ways:

- by boiling water to produce steam to drive turbines
- by using arrays of photovoltaic (PV) cells

Ideal conditions are found in desert regions in the tropics and sub-tropics, where skies are clear for 300 or more days a year and solar radiation is intense. Invariably, the areas of highest potential are sparsely populated and so there are few objectors. Although the development of solar power is in its infancy, the potential is huge. For example, an area the size of Portugal in the Sahara Desert could produce electricity equal to the combined output of all the world's power stations. To date, major investment has been limited to southern Spain and California's Mojave Desert.

Geothermal energy

Geothermal power stations, which tap the heat of the Earth's interior, have been developed in Iceland, New Zealand and California. Like solar energy, the potential is huge and the energy supply is inexhaustible and non-polluting. The UK's first commercial geothermal energy power station is planned for Cornwall, where hot rocks (granite) lie just 4 km below the surface. Water injected into a bore hole will be heated to 150°C and used to drive turbines. The geothermal power plant will generate enough electricity to supply the Eden Project and meet the electricity demands of 5,000 households.

There are few objections to geothermal energy on environmental grounds. The environmental footprint of geothermal stations is small and surplus warm water can be used for heating and for recreation and leisure (e.g. Iceland's Blue Lagoon).

4.3 Nuclear power

Nuclear accidents at Three Mile Island in Pennsylvania in 1979 and Chernobyl in 1986 heightened public awareness of the dangers of nuclear energy and hardened opinion against the nuclear option. For over 20 years no new nuclear power stations were planned in either the USA or the UK. Even today there remain strong arguments against nuclear power:

- accidents involving nuclear power are potentially lethal. They could kill and injure thousands of people and cause long-term health problems for many more
- any radioactive particles released into the environment remain toxic for thousands of years
- no secure long-term repository for radioactive waste and spent uranium fuel has yet been built anywhere in the world
- nuclear installations are potential targets for terrorist attacks
- taking account of construction and decommissioning costs, electricity generated by nuclear power is more expensive than electricity generated by burning coal and other fossil fuels

Despite these objections and public scepticism, several governments in developed countries are planning to increase their new nuclear capacity. Nuclear power is once again on the agenda because:

- it creates no carbon emissions and helps countries achieve the targets set by the Kyoto Protocol
- it is more sophisticated and safer than it was 20 or 30 years ago
- as polluting coal-fired generating capacity is phased out, renewables like wind power, in the short term, do not have the potential to fill the so-called 'energy gap'. Nuclear energy offers the only practical alternative
- it provides greater energy security for developed countries because most uranium resources are in stable and friendly countries such as Canada and Australia

1 Defining tourism

The World Tourism Organization defines tourism as 'the activities of persons travelling to, and staying in, places outside their usual environment for leisure, business and other purposes'

2 The environmental impact of tourism

2.1 Mass tourism and ecotourism

Tourism often has an adverse impact on the physical environment, degrading the very resources on which it depends. Damage results from pollution and the overuse and misuse of resources, together with poor management and planning control. Tourism that permanently degrades the environment and reduces its value for future generations is said to be **unsustainable**.

Mass tourism — which caters for large numbers of visitors, is highly concentrated geographically and based around natural resources such as beaches and mountains — is most damaging. In the rush to mass tourism in the 1960s in southern Spain, large stretches of the coastline were degraded by unregulated building, pollution and the destruction of wildlife habitats.

An alternative, more sustainable tourism has become popular in the past 30 years. This is known as **green tourism** or **ecotourism**. It describes small-scale tourism based around special interests such as trekking, wildlife and cultural/historical features. Its environmental impact is far less than mass tourism (even though it often uses the same infrastructure) and it aims to be sustainable.

2.2 Tourism in fragile environments

In the past 30 years improvements in transport, personal mobility and disposable incomes have opened up many new tourism destinations. Tourist operators have also developed the market for more active holidays in exotic locations such as trekking in the Himalayas and the Andes, scuba diving on the Great Barrier Reef, sight-seeing in Patagonia and wildlife safaris in east Africa.

CASE STUDY 1	Tourism in the Arches National Park, Utah
Background	The Arches National Park in southern Utah, USA (Figure 15.1) is a high desert environment on the Colorado Plateau. Rainfall averages just 230 mm a year and vegetation cover is sparse. The Arches is a small (300 km²) but popular park designated in 1971. Its main attraction is the highest concentration of natural stone arches in the world. In 2006, 833,000 people visited the park, the vast majority by car.
Environmental impact	The main environmental problem is damage to the cryptobiotic crust caused by trampling and bikers. The crust, which comprises communities of cyanobacteria, green algae, lichen, fungi and mosses is thin, brittle and extremely fragile. However, it is crucial to the ecology of hot arid and semi-arid environments, protecting soils from wind and water erosion, retaining moisture from rain, inputing nutrients to soils and providing a nursery for seedlings. Once damaged, the crust may never recover. Food chains that depend on the crust will collapse and ultimately soil erosion will result in widespread land degradation.
Responses	The National Park Service provides information to visitors at its visitor centre about the importance of fragile cryptobiotic crusts. Visitors are requested to keep to footpaths and trails, and parking along verges, which often damages vegetation and soils, is banned.

However, many of these exotic destinations, such as high mountains and hot arid and semi-arid environments, support fragile ecosystems and habitats. In these places, where biodiversity and rates of primary production are low, even small disturbances can cause permanent damage to the environment.

Therefore, sustainable tourism in fragile environments has to be within the limits of the environment's **carrying capacity**. Sadly, even with greater environmental awareness and the growth of ecotourism, many environments have been degraded by visitor pressure.

3 Sustainable tourism

The World Commission on Environment and Development has defined **sustainable tourism** as 'development that meets the needs of the present without compromising the ability of future generations to meet their own needs'.

The uncontrolled growth of mass tourism in southern Europe from the 1960s to the 1980s exemplifies the problems of unsustainable tourism.

Summer drought in southern Europe puts huge pressure on limited water resources at a time when demand, inflated by tourism, reaches its peak. Excessive demand has led to the overpumping of groundwater, resulting in falling water tables, incursions of sea water and the depletion of water resources in Mallorca and parts of Andalucia.

Meanwhile, the physical growth of resorts has caused a loss of habitat. In Mallorca areas of heathlands, pine forests and wetlands have declined as the island's tourism industry has boomed.

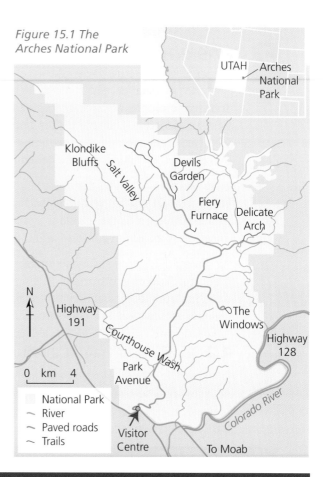

Figure 15.1 The Arches National Park

CASE STUDY 2 Sustainable tourism in Costa Rica

Costa Rica is a small country in Central America with a population of 4.25 million (2009). Tourism developed in Costa Rica only in the 1980s. The country's mosaic of exotic natural environments — volcanoes, rainforest, cloud forest, high plateaux — are its principal tourism attractions. Visitors come to Costa Rica to experience its diverse scenery and wildlife (ecotourism) and for activities such as fishing, surfing, wind surfing and mountain biking. One-quarter of the territory comprises protected conservation areas, including National Parks.

Costa Rica's government has pioneered a sustainable approach to tourism. It promotes limited development that aims to respect the environment and culture of local people. Hotels are small (only 5% have more than 100 rooms) and a significant number are owned and operated by Costa Ricans themselves rather than being international chains.

So far, tourism has not degraded the environment or local cultures and has brought significant benefits to native people. Tourism has become the leading sector in the economy and in 2007 there were 1.9 million visitors. In 2007 it accounted for 13% of total employment, had a turnover of nearly US$2 billion and generated just over 8% of GDP.

Sustainable tourism has to balance economic growth with conservation of the environment. Ideally, sustainable tourism should:
- maintain the quality of the environment on which tourism depends
- provide a high-quality experience for visitors
- improve the quality of life of the host community

4 *Management and planning for sustainable tourism*

There is a clear need to manage and plan for sustainable tourism that conserves environmental resources, protects native cultures and brings benefits to local people.

4.1 National Parks in England and Wales

In the past 60 years, the National Parks of England and Wales have shown how it is possible to balance environmental protection and economic development with a sustainable tourism industry.

The National Park movement began in the USA in the second half of the nineteenth century. The world's first National Park was Yellowstone in Wyoming, established in 1872. By comparison, National Parks made a late start in England and Wales. It was not until 1949 that the British government introduced the legislation that created them.

The National Parks and Countryside Act 1949 designated ten National Parks in England and Wales (Figure 15.2). These parks covered the areas of the highest environmental quality in England and Wales. All ten had been established by the end of the 1950s. Subsequently, a further three parks have been created: the New Forest, the Norfolk Broads and the South Downs. Since 2000, two National Parks have been designated in Scotland.

Figure 15.2 National Parks in the UK

Although the primary aim of National Parks is conservation, the 1949 Act also stated that they should be accessible for the enjoyment of the public through quiet recreation. In other words, National Parks also have a recreation and tourism function.

The main attraction of the parks for recreation and tourism are their landscapes, habitats and wildlife. However, if recreation and tourism are to be sustainable, these activities must not damage or degrade natural resources. The task of managing each park — ensuring its sustainable use and resolving conflicts between conservation, tourism and development — rests with the National Park Authorities (NPAs). Each National Park has its own NPA made up of local councillors, parish councillors and members nominated by the Secretary of State for the Department for Environment, Food and Rural Affairs (Defra). Where conflict arises, conservation has priority. This is known as the **Sandford Principle**.

The overall success of **sustainable management** in the National Parks of England and Wales can be gauged if we compare today's landscapes with those of 60 years ago. National Parks in England and Wales have proved hugely popular, receiving over 100 million visitors a year.

The Yorkshire Dales National Park alone received 9 million visitors in 2008. Despite the pressure that comes with such popularity, the parks have successfully conserved their landscapes, habitats and wildlife. In the Dales, for example, spectacular landscapes such as Upper Wharfedale remain largely unchanged. Sustainable management has ensured that their beauty is unimpaired and can be enjoyed by future generations.

CASE STUDY 3   Sustainable management in the Lake District National Park

Background	The Lake District National Park (LDNP) was established in 1951 and is the largest park in England and Wales (Figure 15.3). It is the only truly mountainous region in England, with deep glacial valleys carved into ancient volcanic rocks, occupied by ribbon lakes. Before its designation as a National Park, the Lake District had been an important centre for tourism since the mid-nineteenth century. Tourism was mainly concentrated in a handful of lakeside settlements such as Windermere, Ambleside, Keswick and Grasmere.
Management structure	The LDNP is managed by a National Park Authority (NPA). The NPA is the sole planning authority in the park and is responsible for the control, development and allocation of land use. Annual funding (£6.6 million) comes from Defra and Cumbria County Council. Parking fees, planning approvals and sales at visitor centres make up one-third of the park's income. Apart from planning control, the NPA also has responsibility for managing footpaths, providing a ranger service and supplying tourist information.
Environmental conflicts	The NPA must address a number of tourism-related issues. These include traffic congestion, changes in farming practice and land use, conflicts between different recreational uses, lake management and footpath erosion. • Traffic congestion is a problem as the larger settlements and major roads leading into the LDNP (e.g. the A591 from the M6) become heavily congested with long queues at peak times. Traffic is getting worse and the current situation is unsustainable. • Changes in farming practice and land use, which affect the landscape. They are largely determined by the EU's Common Agricultural Policy (CAP) and various environmental schemes. Lower levels of stocking on the fells to protect the environment could threaten the Lake District's open landscape. The use of agro-chemicals, including pesticides, destroys habitats for wildlife. Forests and woodlands need careful management and control. In the past, the extensive coniferous planting for commercial timber production has damaged landscapes in areas such as Ennerdale and Grizedale. • Conflict between different recreational users occurs on the larger lakes and elsewhere in the park. Motorised water sports conflict with quieter forms of recreation (e.g. sailing, angling, bird watching). Mass tourism conflicts with activities such as hill walking, mountain biking and climbing. • Lake management is the responsibility of the NPA, which manages 16 major lakes and smaller water bodies. The more popular lakes are vulnerable to tourism pressure. A particular problem has been the pollution of Windermere (due to increased phosphorus loads in treated sewage effluent discharged into the lake and runoff of agro-fertilisers) and subsequent eutrophication, threatening trout, salmon and arctic char populations. • Footpath erosion occurs — there are nearly 3,000 km of footpaths and bridleways in the park. Hill walking is a popular activity and footpaths in the more accessible areas (e.g. Langdale, Coniston, Kentmere) have suffered severe erosion.
Management responses	• The NPA has proposed solutions to traffic congestion problems, including the introduction of shuttle buses, congestion charges for entry to the park, raising parking fees and promoting more sustainable transport such as cycling, walking and horse riding. • Farmers are encouraged to use traditional and more sustainable methods through environmental schemes supported by the EU (e.g. cross compliance) and Defra (e.g. Environmental Stewardship, Farm Woodland Premium). Conifer plantations such as Grizedale and Whinlatter have been developed for multiple use (e.g. biking, nature watching). • Conflicts between recreational users have been resolved by zoning. This approach reserves particular parts of the park for different types of recreation. The busier central valleys cater for mass tourism, with restrictions placed on traffic congestion, noise and visual pollution. The bulk of the park (including the fells) is reserved for quieter recreation. There, management is less intrusive and land use is less intensive. Conflict between recreational users on Windermere has been resolved by imposing a 16 km/h speed limit on the lake. • Lake pollution can be tackled by the NPA by restricting new housing development around lakes such as Windermere. The Environment Agency will improve sewage treatment capacity and address the issue of runoff from farmland of agro-fertilisers. • Footpath restoration schemes, funded by the NPA, National Trust, Natural England and the National Lottery, are in operation in the most popular hill areas. Tens of kilometres of eroded paths have been restored by draining and building hard stone surfaces.

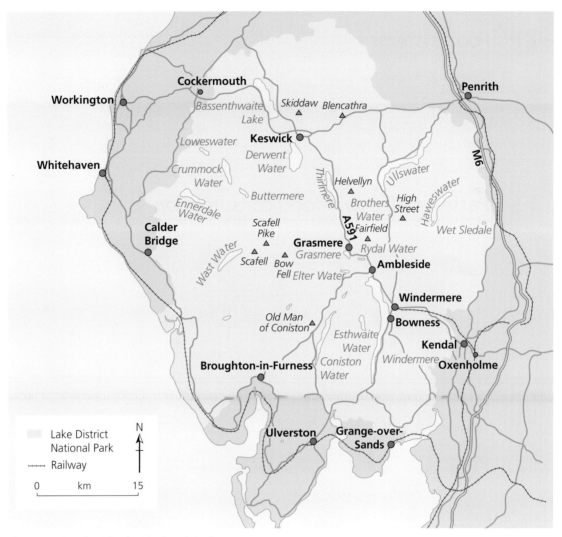

Figure 15.3 Lake District National Park

1 Food systems

Food systems consist of chains of people — food producers, food processors, food retailers and restaurants — and are the means by which urban societies get their food supplies. The food system idea is most relevant to developed countries and to commercial farming, which is geared to market demand. For non-commercial farming and **subsistence** and **semi-subsistence** societies, food systems have less relevance. Even so, a large proportion of food produced in developing countries is marketed, though food channels may extend no further than the nearest village market.

2 Patterns of global food consumption

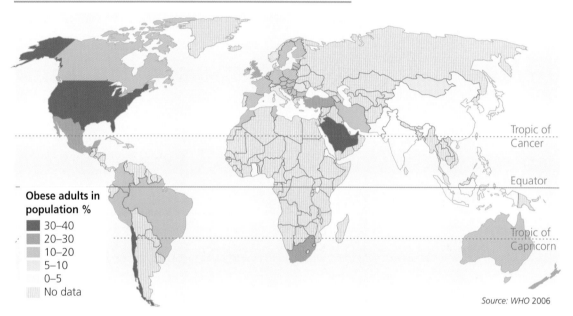

Obese adults in population %
- 30–40
- 20–30
- 10–20
- 5–10
- 0–5
- No data

Source: WHO 2006

Figure 16.1 Global obesity

At the global scale, spatial patterns of food consumption are highly uneven. At one extreme, there are problems of overeating and **obesity** in many developed countries (Figure 16.1). At the other extreme

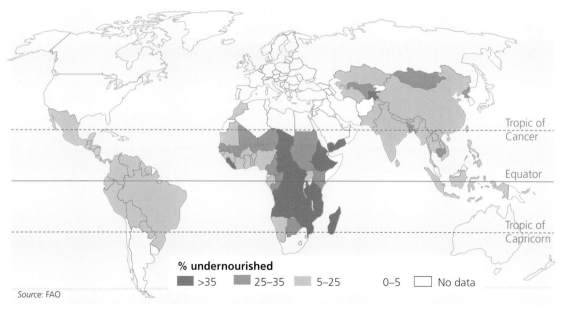

% undernourished   >35   25–35   5–25   0–5   No data

Source: FAO

Figure 16.2 Global undernutrition

in some parts of the developing world, and especially in sub-Saharan Africa, the problems are food shortages, **undernutrition** and **malnutrition** (Figure 16.2). **Undernourishment** refers to the condition of people whose dietary energy intake is below a minimum requirement for maintaining a healthy life.

Table 16.1 shows that of the ten countries with the highest incidence of undernourishment, nine are in sub-Saharan Africa. Meanwhile, nearly 30% of the world's population suffers from some form of malnutrition caused by an unbalanced diet and a lack of particular nutrients. Malnutrition is widespread in the developing world, although is not unusual among the lowest income groups in developed countries.

Table 16.1 Prevalence of undernutrition, 2005

Country	% undernourished	Country	% undernourished
Dem. Rep. Congo	76	Angola	46
Eritrea	68	Ethiopia	46
Burundi	63	Zambia	45
Haiti	58	Central African Rep.	43
Sierra Leone	47	Zimbabwe	40

In contrast to the incidence of undernutrition and malnutrition, the top ten countries with the highest levels of nutrition are all in the developed world (Table 16.2). Overnutrition causes obesity and is associated with health problems such as diabetes, heart disease and cancer.

Table 16.2 Prevalence of overnutrition, 2005

Country	Average Kcal per person per day	Country	Average Kcal per person per day
USA	3,830	Italy	3,680
Luxembourg	3,780	Israel	3,610
Belgium	3,700	France	3,590
Greece	3,690	Canada	3,560
Ireland	3,680	Germany	3,510

3 *Food security*

3.1 Food security and food availability

Food security means that people have assured access to sufficient food to lead a healthy life. However, food security is not the same as **food availability**. Food can only be bought if it is sold at prices people can afford. Food may be available in a country, but without adequate transport, storage facilities and government cooperation people may not have access to it. This was the case in Myanmar (Burma) following the Cyclone Nargis disaster in 2008. The Myanmar government refused to allow ten helicopters from the UN World Food Programme to fly relief missions, and US warships, loaded with emergency food supplies, were not allowed to offload their cargoes.

Famines

Famines are widespread food shortages that lead to a sharp rise in mortality. However, the concept of famine is not as simple as it might appear. For example, famines are often highly localised and may affect only one social or economic group. Famines are also not always the result of an absolute shortage of food. They can occur when a breakdown in the marketing system occurs or when people simply cannot afford to buy food that is available. In famines, starvation is rarely the cause of death. Most often, undernutrition reduces the body's resistance to infection and people die from a range of diseases.

Famines can have several causes:

- severe reduction in food production due to harvest failure (e.g. drought in the Horn of Africa in the mid-1990s, flooding in Somalia in 2006)
- political instability that disrupts food production (e.g. China's Great Leap Forward in the early 1960s, where up to 3 million people died from famine, recent civil wars in Africa such as in Sudan)
- overexploitation of declining soil and water resources in dryland environments due to rising population levels (e.g. Niger in 2005)
- natural disasters in developing countries such as tropical cyclones that destroy crops (e.g. floods in Bangladesh in 1974, Cyclone Nargis in Myanmar in 2008)

Poverty and famine

The economist Amartya Sen has argued that famines are not always caused by a decline in food availability. Some famines result from a deterioration in the entitlements of the most vulnerable members of society. In other words, poor people have limited access to food because of their weak purchasing and bargaining power.

A person's entitlements might include land ownership and other assets, occupation and status. However, for most people **exchange entitlements** are most important. This means having enough income to purchase food and prevent undernutrition. People who depend wholly on exchange entitlements are most at risk in times of food shortage when food prices rise steeply. They have no other entitlements to fall back on to help them survive the crisis. As a result, famines inevitably hit the poor harder than any other group (Table 16.3).

Table 16.3 Household characteristics and entitlements

Status/income	Land	Other assets	Savings/stored food
Farmer: above average income	Owner-occupier of 10 ha	Livestock, oxen, ploughs and other means of production	Significant savings
Smallholder: average income	Tenant; fixed rent; farms 2 ha	Owns a few livestock. All means of production owned by landlord	Small savings
Landless labourer: below average income	None	None	None

4 Food supplies and population growth

There are two opposing theories of food supplies and population growth: the pessimistic view first proposed by Thomas Malthus and the optimistic view associated with Esther Boserup.

4.1 Malthusian theory

In 1798 Thomas Malthus argued that population pressure was the direct cause of famine. He assumed that food supplies could only increase at an arithmetic rate, while population growth increased geometrically (Figure 16.3). As a result, according to this view, population growth would inevitably outstrip the growth of food supplies, causing famine, disease, war and massive mortality. The only way to avoid such a catastrophe was to control fertility.

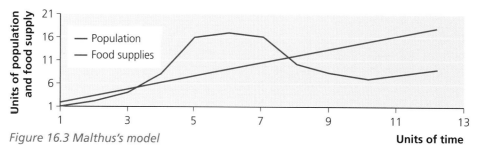

Figure 16.3 Malthus's model

4 Food supplies and population growth

Although disaster on the scale predicted by Malthus did not occur, Malthus's ideas have remained popular among some economists. The so-called neo-Malthusians argue that many famines in the past 30 years, particularly in sub-Saharan Africa, have their origin in excessive population growth.

4.2 Boserup's theory

In 1965 Esther Boserup presented an alternative view to Malthus (Figure 16.4). She argued that population growth would not necessarily lead to food shortages. Rather, increased demand for food and rising prices would give farmers an incentive to increase output. This could be achieved in a number of ways:
- by working longer hours
- by cultivating more land
- by innovation, which might include the development and application of new technologies (e.g. genetically modified (GM) crops, breeding new crop strains) and intensifying output by multi-cropping or irrigation

The history of world population growth and food production in the past 50 years lends support to Boserup's theory (Figure 16.5).

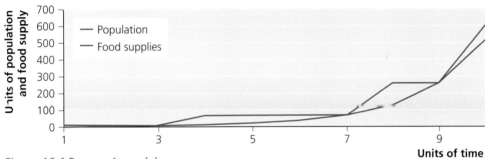

Figure 16.4 Boserup's model

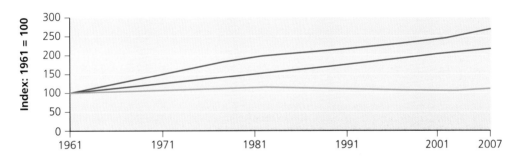

Figure 16.5 World population growth, cereal production and cereal area, 1961–2007

Box 1 *Population growth and global food resources*

The protagonists fall into two camps: the environmental pessimists and the technological optimists. The pessimists argue that the world's population will increase from just over 6 billion in 2000 to 9.5 billion by 2050. Global demand for food is expected to double in this period. Feeding this huge increase in numbers will mean, among other things, intensifying farming and converting forest to farmland, which will hasten land degradation and accelerate habitat destruction and climate change. Despite big advances in total food production in the past 40 years, per capita food consumption has barely increased in some parts of the world (e.g. no improvement in west Africa between 1996 and 2005). It is estimated that global cropland production is 13% lower than would have been the case without soil degradation and overexploitation of resources. In the long term, sustainable growth of food production is impossible. The UN estimates that already nearly 1 billion people do not get enough to eat and that 25,000 people a year die from hunger and related causes.

The optimists paint a more positive picture. They believe that technological innovation such as GM crops will ensure food supplies in future. They also point out that global food production has increased faster than population in the past 50 years and that in recent years increases in demand for food have been met by intensification, mechanisation, technological advance and irrigation. Some experts believe that the global amount of potential cultivable arable land is double that currently being used. Only 38% of the Earth's land area is used for farming and, of the area farmed, only 29% is used to grow arable crops. Huge quantities of food are wasted in harvesting, transporting, storage, distribution and retailing. Of the 100–130 million tonnes of fish caught annually, 30 million tonnes is discarded (Figure 16.6) . Meanwhile, the practice of feeding cereals and other crops to livestock is extremely wasteful. Abandoning this practice could feed another 3.5 billion people (FAO estimate).

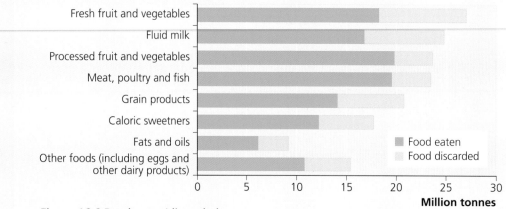

Figure 16.6 Food eaten/discarded

CASE STUDY 1   Famine in Niger, 2005

Background	In the summer of 2005, 3 million people in Niger suffered severe food shortages and famine. Thousands died from starvation and disease in the southern provinces of Maradi and Zinder.
Drought	Before the onset of famine, southern Niger had experienced 13 months of drought. Because of drought, subsistence crops of millet, sorghum and maize failed. In December 2004 the UN's Food and Agriculture Organization (FAO) said that drought and desert locusts were responsible for the poor harvests and food shortages.
Population growth and land degradation	Food shortages in southern Niger are not just caused by drought. Niger's population doubled between 1985 and 2005, putting enormous pressure on already limited resources of water, soils, woodland and wildlife. Driven by population pressure and more mouths to feed, the drier wooded areas have been deforested by overgrazing and deliberate clearance. Without a protective cover of trees, soils are blown away. Crop yields have declined in the more humid areas as farmers overcultivate the land and clear trees which conserve the soil and protect it from erosion.
Carrying capacity	Carrying capacity is the maximum sustainable population in an area that can be supported at a certain standard of living. Rapid population growth in Niger's hot semi-arid environment has exceeded carrying capacity. The result is a classic Malthusian situation: rising demand for food, diminishing resource base and reduced food resources per capita.
Response	Emergency aid including food, medical supplies, aid workers and money was provided by individual governments, the UN and various charities. Food, water and medicines were available at emergency relief centres. Once the immediate crisis was over, aid agencies focused on long-term problems such as poverty, land degradation and the sustainable management of soil, water and woodlands. Intercropping — planting trees and shrubs between cereals — was promoted. It conserves the soil, prevents erosion and is a source of fuelwood. Pastoral farmers are encouraged to keep sheep and goats, which are more drought tolerant than cattle. More efficient systems were developed for collecting and storing rainwater and providing additional water supplies from wells.

5 *Increasing food production: the technological 'fix'*

Increasing local food production is the long-term solution to problems of food shortages. Technology can play a key part in this. Well-known examples of new technologies being used to increase food production are the Green Revolution and genetically modified (GM) crops.

Increasing food production: the technological 'fix'

5.1 The Green Revolution

The **Green Revolution** of the 1960s and 1970s was based on a package of new **high-yielding varieties** (HYVs) of rice and wheat, irrigation and chemical fertilisers. The primary objective of the Green Revolution was to increase crop yields and therefore food supplies in the developing world. In Asia, where the Green Revolution was widely adopted, food production increased substantially. For example, India's production of padi rice increased from 53.5 million tonnes in 1961 to 80.3 million tonnes in 1980.

Despite successfully increasing total food supply, the Green Revolution was variable in its impact. Higher income groups in rural societies, with access to irrigation and the resources to purchase chemical fertilisers, gained most. The new HYVs also had little immunity to disease so that farmers, if they were to benefit from the new crops, had to spend more money on pesticides. Not surprisingly, the poorest groups (landless families, general labourers, small-scale farmers) gained few benefits. The new technologies of the Green Revolution tended to increase inequalities in rural societies and did little to improve food security for the poor. Therefore, undernutrition and poverty are still widespread.

The Green Revolution also had a negative effect on women in some rural areas. The need for cash income to pay for fertilisers and pesticides forced many women to work as agricultural labourers. This process forced down rural wages so that millions of women had insufficient income to improve their diets and take advantage of the increased production of rice and wheat.

5.2 Genetically modified crops

The HYVs of the Green Revolution were produced by selective plant breeding for bigger and better crops. **Genetically modified (GM) crops** take selective breeding a stage further by transplanting desirable genes from one plant into another. GM has enormous potential to solve the world's food shortages, but it is highly controversial. By 2009, after 13 years of use, only 25 countries had adopted some GM crops. Currently, seven EU countries grow GM crops but they occupy only 108,000 ha of farmland out of the 60 million ha available in the EU. The only GM crop grown in the EU in 2009 was insect-resistant maize.

The supporters of GM crops (especially the biotechnology companies involved in their production) argue that they are safe and will deliver significant increases in global food production, especially in the developing world. The specific advantages of new GM crops are:
- varieties can be developed that are drought tolerant, saline tolerant (large areas of farmland in semi-arid environments have become too saline for traditional crops), insect resistant, higher in nutritional value and more productive
- they require lower inputs of pesticides (good for the environment and cheaper for farmers)
- some varieties can be grown with little tillage and ploughing, which reduces the threat of soil erosion
- by increasing production, the need to bring new land into cultivation is reduced; this again is good for wildlife and the environment

Despite these advantages, the relatively low take-up of GM crops indicates concerns about their long-term impact on human health, the environment and ecosystems. Many governments in the developed world, aware of public disquiet, are proceeding with caution. Possible disadvantages of GM crops include:
- their unknown impact on food chains and natural ecosystems
- the possibility of their 'escape' and potential capacity to grow more vigorously than other crops and wild plants
- cross-pollination with other plants to produce uncontrollable weed species
- genes from the genetically modified plants could be transferred to the insect pests so that they become resistant to pesticides
- when consumed, foods that have antibiotic resistance by gene technology may give the same property to people, with huge implications for human health
- transferred genes may contaminate other organisms undesirably, causing a biological disaster

6

Water supply

6.1 Water budget

Within the natural **hydrological cycle** (Figure 16.7), the water budget determines the potential water resources available for human use.

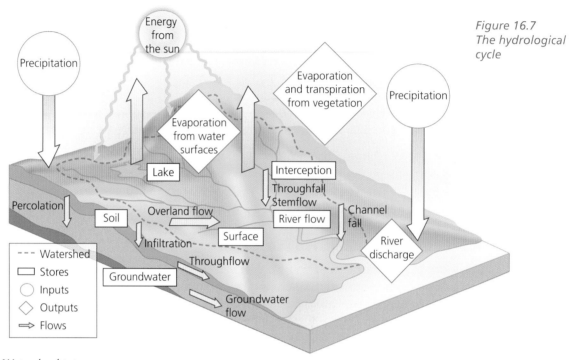

*Figure 16.7
The hydrological
cycle*

Water budget:

Precipitation = evapotranspiration + runoff + soil moisture + surface water + ground water

Because of **evapotranspiration**, only a fraction of the precipitation falling in an area is normally available for human use. That part of precipitation which is not lost to evapotranspiration either becomes **runoff** or it enters storage in the soil and permeable rocks. Human exploitation of water resources focuses on surface runoff and **groundwater** stores.

6.2 Physical factors influencing water supply

In most parts of the world, regional climate conditions are the main influence on water supply. Areas with high mean annual precipitation all year round usually have abundant water resources (e.g. Norway, New Zealand's South Island). However, where precipitation is seasonal (e.g. the Mediterranean, the African savannas and monsoon Asia), acute water shortages are common during the dry season. Moreover, a larger amount of precipitation is lost to evaporation in the tropics and sub-tropics compared to higher latitudinal areas. The **inter-annual variability** of precipitation also affects water availability. In most semi-arid environments, inter-annual variability is higher than in more humid environments. As a result, drought is a common occurrence across much of Australia and in the Sahel region of Africa.

At a smaller scale, other physical factors influence water supply. In the chalk regions of southern England and northern France, surface water is scarce and for centuries has determined the distribution of settlement. Rivers that derive their flow from distant humid areas (e.g. the Nile in central Africa, the Colorado in the Rockies) provide arid environments like Egypt and Arizona with water supplies.

At the global scale, the world's most water-rich continent is South America (Table 16.4). Demand is low in this region, so per capita water resources are high. Asia is also water-rich but its huge population creates a high demand and water scarcity in countries such as India and China. Europe's supply and demand situation is similar to Asia's.

Table 16.4 The world's renewable water reserves

Region	Water reserves (km³/year)	Annual usage (km³/year)	Population (millions)
Africa	5,723	213	996
North and Central America	7,621	622	534
South America	17,140	165	394
Asia	14,872	2,131	4,102
Europe	2,783	384	521
Former Soviet Union	5,287	251	283
Oceania	1,669	26	35

6.3 Human factors influencing water supplies

Human activities can both increase and decrease the quantity and quality of water supplies. Surplus water in rivers, which might otherwise be lost in runoff, can be stored by damming valleys and creating reservoirs. For example, the entire annual floodwaters of the Colorado River are stored in Lake Powell in Utah. Surplus water can also be stored (though with more difficulty) in permeable rocks such as chalk and sandstone by **artificial recharge**.

Human activities often have a negative effect on water resources. The most obvious example of this is pollution. Rivers can be polluted by untreated sewage, industrial effluent and runoff from chemical fertilisers and pesticides from farmland. This kind of pollution may be routine (this is increasingly rare in developed countries but remains common in developing countries) or accidental (e.g. the pollution of the River Rhine at Basel in 1985 following a spillage of toxic chemicals). Pollution is sometimes so severe that water cannot be abstracted for public water supply or crop irrigation (e.g. Lower Yangzte River). Overpumping of groundwater in coastal zones can also cause a loss of water resources, allowing incursions of salt and brackish water into freshwater aquifers (e.g. Bangladesh, Nile delta).

7   *Water demand*

In the past 50 years global population growth and rising standards of living have greatly increased the demand for water. Even so, less than 7% of the world's annual renewable water resources are currently utilised. There is, however, considerable variation in the use of available resources between countries. In India and Spain more than one-third of annual renewable water resources are used. In countries such as Israel, Jordan and the United Arab Emirates (UAE), consumption exceeds rates of annual water renewal. This situation is only sustainable if water is imported or, as in the case of the UAE, sea water is desalinated.

Freshwater is a finite resource but demand, driven by population growth and rising standards of living, rises inexorably. According to the World Bank, some 1.1 billion people in developing countries have inadequate access to water and about 700 million in 43 countries live below the water stress threshold of 1,700 m³ per person per year. The mismatch between supply and demand is most acute in regions of high population density with semi-arid climates such as sub-Saharan Africa, the Middle East, India and parts of China (Figure 16.8). In these regions supply is barely sufficient to meet demand, placing in doubt the sustainability of population and economic growth. In the southwest USA the rapid growth of cities such as Phoenix, Tucson and Las Vegas since the 1970s has sent the demand for water soaring. However, there has been no comparable expansion of water resources. Water from the Colorado River is now fully utilised and Las Vegas wants to exploit groundwater resources in the Mojave Desert. However, the plan is controversial, not least because groundwater abstraction and any resulting fall in the water table could destroy a number of oases and their unique wildlife. Water shortages due to rising demand threaten the future of economic development through the southwest region of the USA.

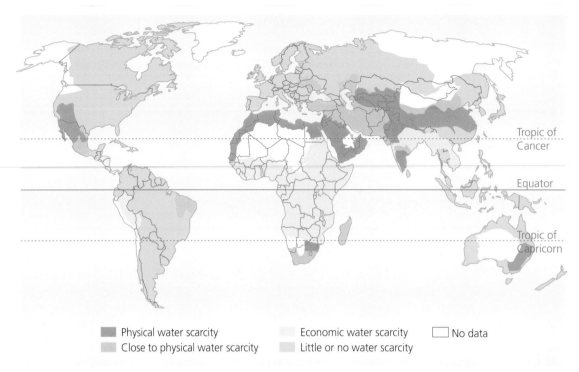

Figure 16.8 Global water scarcity

- ■ Physical water scarcity
- ■ Close to physical water scarcity
- ░ Economic water scarcity
- ▓ Little or no water scarcity
- □ No data

CASE STUDY 2   Drought in the Murray-Darling Basin, Australia

Background	The Murray-Darling river system drains one-seventh of Australia (Figure 16.9). The basin includes four states: Queensland, New South Wales (NSW), Victoria and South Australia. The source of the Murray is the Snowy Mountains of NSW. Most of the drainage basin is either arid or semi-arid. *Figure 16.9 Murray-Darling basin*
Water resources	Water resources declined steeply following a 7-year drought (2009), the worst for over 100 years. The increased frequency of drought and higher temperatures could be due to climate change and global warming. Two-thirds of all the water used for irrigation in Australia comes from the Murray-Darling basin. Adelaide, one of Australia's largest cities, relies on water from this source. Current demand for water exceeds supply. The problem is one of unsustainable use brought into focus by recent severe drought.
Impact	In dryland environments, cultivation depends on irrigation water. Levels of water use in southern and eastern Australia are unsustainable; the region is running out of water. Water allocations to irrigation farmers have been cut or suspended. Ecosystems are dying and the Murray-Darling basin has already lost 90% of its wetlands and 80% of its bird population. Complex governance of the river system that covers four states has led to excessive water allocations and wastage. The Murray-Darling Basin Authority, which is responsible for water resources, sets limits on water extraction. The federal government will spend US$330 million buying water entitlements from farmers in Queensland and NSW and re-leasing them back to the river.

8   *Development of water resources and its impact*

Water resource development often has widespread impacts on the environment, human health and economic activity.

8.1 Environmental impacts

Dams have a number of damaging effects on the physical environment. They:
- flood valleys, destroying habitats and wildlife
- change microclimates around reservoirs
- accelerate erosion and mass movements on valley slopes above the reservoir
- modify water temperature and water turbidity downstream
- prevent flooding downstream

The last dam to be built on the Colorado River was at Glen Canyon. Completed in 1963, the dam had two main functions: to generate around 1,000 MW of hydroelectric power and to provide water supplies to the fast-growing cities and irrigation agriculture in the parched southwest USA. From the start the dam was controversial. It was strongly opposed by conservationists because of its adverse environmental effects. Thousands of square kilometres of superb scenery were flooded by the new reservoir, Lake Powell. The reservoir was big enough to store the entire annual flood on the Colorado River, drastically changing the river's regime. Downstream, the absence of floods and sediment led to the erosion of sandbars — important habitats for fish, amphibians and birds. Meanwhile, indigenous fish species, unable to tolerate the cold clear water, became extinct. The absence of floodwaters, which before the construction of the dam cleared out bankside vegetation, allowed the invasion of alien species. Tamarisk shrubs now dominate the bankside vegetation, creating new habitats and ecosystems.

8.2 Impacts on human health

The rapid expansion of Los Angeles in the late nineteenth century created water shortages for the city. In response the Los Angeles Department of Water and Power acquired most of the land and its water rights in the Owens Valley, an arid region between the Sierra Nevada and Death Valley, 400 km to the east. It constructed the Los Angeles aqueduct to re-route the Owens River — the primary water source for Owens Lake — to Los Angeles. By 1924 Owens Lake had dried up completely. Today, the lake bed is the largest stationary source of pollution in USA. Wind-blown alkaline dust containing toxic metals exceeds air quality standard for particulates (PM10) 20 to 30 times a year, creating respiratory problems for residents in Ridgecrest, Lone Pine and neighbouring settlements.

8.3 Impacts on economic activities

The Aral Sea in what was previously Soviet central Asia was the scene of a disastrous experiment in water resource development from the 1960s to the 1980s (Figure 16.10). The cause of the disaster was the diversion of the Aral Sea's two main feeder rivers, the Amu Darya and the Syr Darya, to supply irrigation water for the commercial production of cotton and rice. The outcome was ruinous. Water levels fell by 17 m as the 56 km³ of water normally delivered by the rivers fell to zero. By 2001 the Aral Sea had shrunk to one-tenth of its original surface area. Before the 1960s the region had a vibrant economy, with thousands of people relying on agriculture and fishing for their livelihoods. However, overirrigation led to waterlogging and salinisation of 40% of irrigated farmland, while pesticides, fertilisers and salt polluted groundwater. Eventually 2 million hectares of fertile farmland were abandoned. The result was land degradation on a grand scale. Meanwhile, the commercial Aral fishery, which had supported 60,000 people and recorded annual sustainable catches of 40,000 tonnes, collapsed. Thousands of people lost their jobs, creating poverty and out-migration. Former coastal villages and towns now lie stranded up 70 km from the shoreline.

However, the effects of the Aral Sea debacle were not just economic. As the sea dried up, the regional climate changed and entire ecosystems (e.g. 500,000 ha of wetlands and marshes) were destroyed.

In addition to economic hardship, local people faced serious health problems (e.g. kidney and liver disease, cancer, bronchitis) caused by polluted drinking water and dust storms that combined a lethal mix of salt particles, pesticides and chemical fertilisers. Today, infant mortality is one of the world's highest.

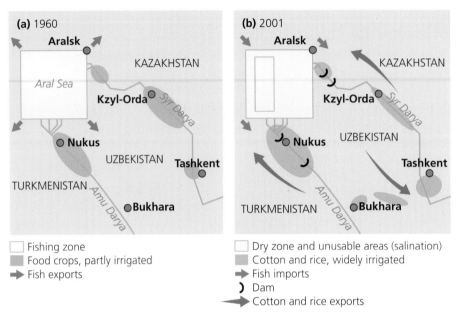

(a) 1960

(b) 2001

Figure 16.10
The Aral Sea: socioeconomic impacts, 1960 and 2001

☐ Fishing zone
■ Food crops, partly irrigated
➡ Fish exports

☐ Dry zone and unusable areas (salination)
■ Cotton and rice, widely irrigated
➡ Fish imports
) Dam
➡ Cotton and rice exports

9 *Sustainable water management*

Sustainable water management involves creating new water supplies, conserving existing supplies and, where possible, reducing consumption. The aim is to guarantee water availability in the long term without depleting water resources and damaging the environment. Table 16.5 describes some methods of water management in the UK.

Table 16.5 Methods of water management in the UK

Method	Description
Increasing supply	
Surface reservoirs	Increase water resources by storing surplus water (e.g. during floods) that would otherwise be discharged to the sea or ocean. The Thames region faces serious water shortages by 2025 (rising demand, drier summers). Expanding supply by building reservoirs is one option. There is a proposal to build a new reservoir in the Upper Thames basin near Abingdon. However, reservoirs are ecologically damaging and unsustainable. Flooding of valleys results in a loss of farmland, habitats and wildlife. The downstream effects may also damage aquatic ecosystems.
Inter-basin water transfers	Aqueducts link river basins, creating in effect a water grid and providing the flexibility to transfer water from areas of surplus to areas of deficit. Some inter-basin transfers already exist; for example, in the northeast of England water from the North Tyne River and Kielder Reservoir can be transferred south to the River Wear and the River Tees, and ultimately into the Yorkshire rivers. Therefore, drought in York could be tackled by importing water from Northumberland through the existing river system.
Artificial groundwater recharge	Surplus rain that falls in the winter months can be stored underground in aquifers (permeable and water-bearing rocks) through artificial recharge. The advantages of groundwater recharge over surface storage are: • no loss of water to evaporation • no loss of land to flooding • less threat of pollution • no overpumping of groundwater Artificial recharge is already used in the Thames basin, where effective precipitation is only 250 mm a year.

Desalination	London's first desalination plant opened at Beckton in 2009. It will produce water for the public supply during times of drought and low rainfall. Output is 140 million litres per day — enough to supply 1 million people. Desalination plants are energy expensive. Those powered by fossil fuels increase carbon emissions and contribute to global warming and climate change.
Reducing demand	
Water metering	Water meters may become compulsory in southeast England in the next decade. At the moment just 30% of households in England and Wales have water meters. When consumers pay for what water they use, they consume 10–15% less. Meters would help to avoid wasteful water use, e.g. taking baths rather than showers, excessive watering of gardens, frequent car washing.
Cutting leaks	It is estimated that one-third of all water entering the mains network is lost through leakages. This is enough to supply 11 million households.
Waste water recycling	Waste water from baths, showers and washing machines, known as grey water, can be collected, treated and reused for toilet flushing (which accounts for one-third of domestic water use). Untreated grey water can be used for garden irrigation. Reedbeds can be used for secondary and tertiary treatment. They are low maintenance, provide additional water resources and are beneficial to wildlife.

10 Transboundary water disputes

There are 263 transboundary river basins (i.e. basins occupying two or more countries) and over 40% of the world's population resides within internationally shared river basins (Figure 16.11). Water is a finite resource, but demand — driven by population growth and economic development — continues to soar. Disputes can arise in river basins where countries share water resources and especially where resources are scarce such as in arid and semi-arid zones. Transboundary water disputes currently occur on the River Jordan and the Tigris-Euphrates in the Middle East, and the Nile in Africa. So far, these disputes have not escalated to military conflict. Solutions to water problems include negotiating protocols and treaties, and giving individual states rights to a share of the water resources sanctioned by international law.

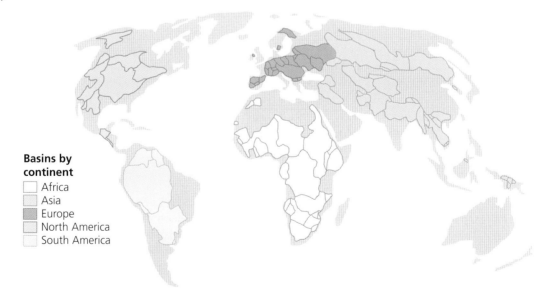

Basins by continent
- Africa
- Asia
- Europe
- North America
- South America

Figure 16.11 Transboundary basins

10.1 Transboundary water resources

Rivers are important sources of water for agriculture, power generation and domestic water supply. Where several countries share a drainage basin, and where water resources are limited in supply, disagreements sometimes occur that can spill over into tension and armed conflict. The most common dispute is where an upstream state unilaterally decides to develop additional water resources, thereby

reducing either the volume of flow or the quality of water to downstream users. As water demand intensifies, the probability of conflict rises, though so far, violent conflict and warfare have been avoided.

Water conflicts on the Nile have been threatened for some time. Ten countries share the Nile drainage basin. Upstream countries such as Uganda, Tanzania and Kenya will need to exploit more Nile water in future for economic development. However, current agreements between the Nile states, completed during the colonial period, did not involve all countries and gave Egypt unfair control over Nile water. Most of Egypt has a hyper-arid climate and, apart from the Nile, the country has few other water resources. Any reduction in flow would be strongly resisted. Moreover, Egypt is militarily the most powerful country in the region and could, if necessary, use armed force to secure its water resources.

10.2 Transboundary water agreements

To avoid transboundary disputes, countries that share an international drainage basin need to agree to the sustainable use and equitable allocations of water resources. There are effective examples of this at a national scale, such as the 1922 Colorado River Compact between the states of the Colorado basin in the USA. However, achieving similar agreements at international level is far more difficult. To date, most transboundary river agreements relate to rights of navigation rather than to water resources.

An exception is the Israeli-Jordanian Peace Treaty signed in 1994. Water is a major issue in the treaty. Both countries recognised each other's rightful allocations to the Jordan and Yarmouck Rivers' water and groundwater in the Arava Valley, and agreed 'not to harm the water resources of the other party' by their own water development projects. Israel's share of water from the Yarmouck was fixed at 25 million m³ per year and allowed a further 20 million m³ per year to be pumped to Lake Tiberias in winter. Jordan's allocation from the Jordan River was 30 million m³ per year. There was also agreement to cooperate on new water projects such as a dam on the Yarmouck and inter-basin water transfers.

CASE STUDY 3   The Tigris-Euphrates drainage basin

Background	The Tigris-Euphrates drainage system covers Turkey, Syria and Iraq. Much of this vast area is an arid, water-poor region. River flows derive from snowmelt and rainfall in mountain headwaters. Ninety per cent of the Euphrates' flow originates in Turkey, the rest in Syria. Syria is responsible for 51% of the Tigris' flow, with 40% from Turkey and 9% from Iraq.
Hydropolitics	Turkey is the upstream state in the basin and Syria and Iraq are the downstream states. This puts Turkey in a strong position because it can control the use of most of the basin's water resources. Iraq, being furthest downstream, is highly dependent on its neighbours. Turkey is also strengthened by its powerful military, which gives it confidence to make unilateral decisions on water usage in the region.
Transboundary issues	The main problem is that the Euphrates rises in Turkey and flows south in Syria and then Iraq. Both Syria and Iraq are highly dependent on the Euphrates for irrigation. Upstream, Turkey's use of Euphrates' water has increased since the mid-1960s by virtue of the Southeast Anatolia Development Project. This is a massive water management scheme involving irrigation networks (covering 1.7 million hectares), 22 dams, 19 HEP plants and costing US$32 billion. The project is aimed at the economic development of the poorest parts of Turkey. When completed, it could reduce downstream water quantity and quality. So far, the development of the basin's water resources has been uncoordinated. On several occasions, flows entering Syria and Iraq have been reduced considerably. In the 1970s, when Syria reduced downstream flows on the Tigris, tension increased and the three protagonists came close to warfare.
Agreements	In 1987 Turkey signed an agreement with Syria to guarantee a minimum flow across the border on the Euphrates of 500 cumecs. In 1992 Turkey refused an Iraqi request to increase flow to 700 cumecs and no agreement was reached. Talks in 1996 stalled because Turkey wanted to allocate water resources by dividing water according to cultivated land, whereas Syria wanted to divide the water equally. By 2003 the issue was still unresolved. Turkey would not sign a treaty to share the water with Syria and Iraq. The lack of internationally recognised water laws help to make this dispute difficult to solve.

TOPIC 17 Pollution and health risks

1 Defining risks to human health

Risks to human health are both short term (**acute**) and long term (**chronic**) and vary over space at global, regional and local scales. Risks also vary historically as society, technology and economic development advance.

1.1 Short-term and long-term health risks

Infectious diseases like influenza (flu) and measles present short-term risks to human health. Disease spreads from person to person up to a peak and then falls off as potential hosts are exhausted. Most people infected recover fairly quickly and acquire immunity to the pathogen. Recent examples of short-term diseases include SARS, bird flu and swine flu.

Even so, short-term diseases can create high levels of mortality. The Spanish flu **pandemic** of 1918–20 infected half a billion people and killed an estimated 50 to 100 million. A pandemic is defined as the geographical spread of a disease to more than two world regions.

Some diseases such as malaria and lymphatic filariasis present long-term threats to human health and are endemic to particular parts of the world. Lymphatic filariasis, for example, affects 120 million people, with nearly 1 billion at risk in 80 different countries. Diseases like these cause long-term illness and have significant economic impact, especially in poor countries. Every year between 300 million and 500 million people contract malaria, out of which nearly 1 million will die.

1.2 Geographical distribution of health risks

At all scales, the geographical distribution of health and disease is often highly concentrated. Globally, populations in rich countries are generally healthier than those in poor countries, reflecting a range of factors such as better hygiene, sanitation, diet and healthcare.

Some diseases are associated with poverty. For example, lack of clean drinking water causes cholera and typhoid, while tuberculosis (TB), which is highly infectious, spreads rapidly in overcrowded slum conditions. Today, these diseases are concentrated in the developing world, where poverty is widespread. However, some diseases, such as cancer, heart disease and diabetes, have much greater incidence in developed countries. They are **diseases of affluence**, related to diet and lifestyle.

Sharp geographical contrasts in health also exist within countries. In cities in developing countries, where poverty is localised in slum housing and informal settlements, rates of ill health and mortality are higher than in the prosperous suburbs. This contrast, though less well defined, is also a feature of cities in developed countries.

Figure 17.1 shows that in inner London the highest rates of infant mortality are concentrated in Southwark, Newham, Haringey and Brent. The rates in Southwark and Newham, two of inner London's poorest boroughs (7 per 1,000) are more than twice that in affluent Richmond (3.1 per 1,000). In fact, only four inner London boroughs have rates of infant mortality below the London average.

Many diseases, because of their specific **epidemiology**, are **endemic** in particular geographical regions. The most obvious ones are tropical diseases spread by parasite vectors (carriers) such as malaria, filariasis and leishmaniasis. Global warming and climate change threaten an expansion of some tropical diseases into mid-latitude environments.

Air pollution creates major health problems in densely populated and heavily industrialised urban areas. Populations suffer above average rates of respiratory illness such as bronchitis, asthma and lung cancer. The causes of ill health include sulphur dioxide, nitrogen oxide, ozone and other toxic chemicals emitted by motor vehicles, power stations and heavy industries.

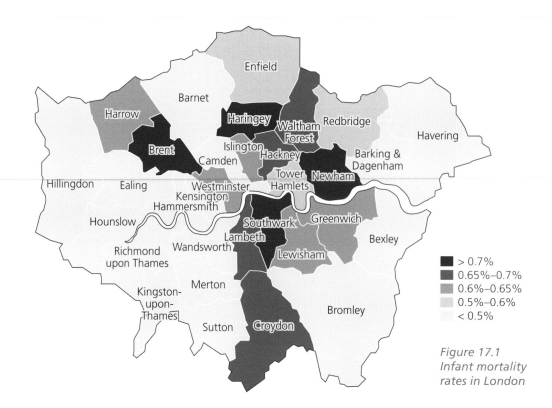

Figure 17.1
Infant mortality
rates in London

Legend:
- > 0.7%
- 0.65%–0.7%
- 0.6%–0.65%
- 0.5%–0.6%
- < 0.5%

1.3 Health-risk patterns over time

In the past two centuries in the developed world average life expectancy at birth has increased from around 40 years to nearly 80 years (Figure 17.2). In the late nineteenth and early twentieth centuries, major advances occurred in the control and understanding of **communicable diseases**:

- immunisation for diphtheria and smallpox
- improved understanding of the epidemiology of disease (e.g. the significance of personal hygiene and clean water to prevent the spread of cholera, dysentery and typhoid)
- mosquito control to fight malaria and yellow fever

Of these methods, only vaccination against smallpox was in use in the first part of the nineteenth century.

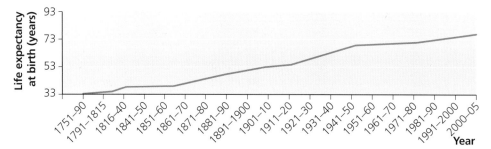

Figure 17.2 Life expectancy in Sweden, 1751–2005

More recent advances in medical technology and disease control have led to progress in treating degenerative diseases of old age, especially cardiovascular disease and cancer. As a consequence, life expectancy has been extended further. Another breakthrough came with the development of penicillin and other antibiotics in the 1940s. In the developing world, the same technologies have contributed to a steep fall in mortality and rising life expectancy.

Advances in human health have also been driven by environmental improvements such as the availability of clean water and proper sanitation. In the late nineteenth and early twentieth centuries in Europe and North America, improvements in sanitation led to huge reductions in child mortality. The

last cholera epidemic in the USA was in 1910–11. In the UK, the decade 1898–1907 saw massive investments in urban sanitation, which reduced infant mortality from 131 per 1,000 to less than 90 per 1,000 (Figure 17.3).

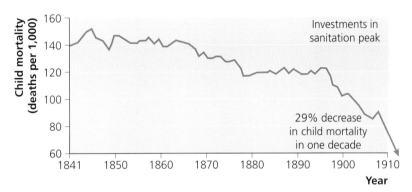

Figure 17.3 UK infant mortality rates, 1841–1910

1.4 Health, quality of life and economic development

Being healthy is defined as a state of physical and mental well-being and the absence of illness. A healthy population is an asset to a state and an important component of human capital. Healthy populations contribute more to economic development because they:

- are more productive — fewer working days are lost to ill health and productivity improves
- allow financial resources that might be devoted to healthcare to be invested elsewhere

For example, a 10% decrease in malaria in a country will, on average, add 0.3% to its economic growth. Malnutrition and its health complications is responsible for a decrease in annual per capita GDP worldwide of between 0.23 and 4.7%.

2 *The causes of health risks*

2.1 The complex causes of health risks

The causes of health risks are complex. They include a combination of physical and socioeconomic factors such as environmental pollution, ecology and climate, poverty and lifestyles.

Environmental pollution

Environmental pollution is linked to a number of diseases and is known to increase health risks. In the UK, leukaemia clusters have for decades been linked to nuclear installations such as nuclear power stations, although causality is hard to prove. Air pollution has several causes. Incinerators are known to emit carcinogenic chemicals such as dioxins and PCBs. Acid rain (caused by sulphur dioxide emissions from coal-fired power stations), nitrogen oxides (from motor vehicle exhausts) and ammonia (from intensive agriculture) can cause premature death from heart disease and lung disorders such as asthma and bronchitis. Polluted drinking water, most often from contamination by sewage, transmits cholera, typhoid and diarrhoeal diseases — major killers in many poor countries.

Ecology and climate

Health risks are greater in the humid tropics than in any other natural environment. High temperatures and high humidity create ideal conditions for bacteria and viruses to flourish. In colder climates the winter season kills most insects such as mosquitoes and flies, which are the main vectors of parasites, bacteria and viruses that cause disease. Tropical diseases such as malaria, yellow fever and dengue fever are transmitted by mosquitoes.

Poverty

The root cause of most ill health, particularly in developing countries, is poverty. This is correlated with:

- malnutrition (unbalanced diet) and undernutrition (too little food), which weaken the immune system and make people more susceptible to disease
- overcrowding, which promotes the transmission of tuberculosis (TB)

- open fires in poorly ventilated houses, increasing the risk of respiratory disease
- contaminated water supplies that transmit water-borne diseases
- lack of education about the links between personal hygiene and the transmission of disease

Lifestyles

In rich countries, lifestyle is a major influence on health risks. In the UK smoking costs the National Health Service (NHS) £5.2 billion a year and accounts for 5.5% of its budget. Smokers' life expectancy is 10 years less than non-smokers and smoking greatly increases the risks of lung cancer, heart disease, emphysema and many other illnesses. Diet-related illnesses cost the NHS £6 billion a year. The main problems are obesity and excess weight caused by a combination of overeating, reliance on diets with high levels of saturated fat, excessive alcohol consumption and lack of exercise. Obesity-related diseases include type 2 diabetes, stroke, some cancers and heart disease. Stressful lifestyles caused by the pressures of work are linked to cardiovascular disease and mental ill health.

2.2 Socioeconomic status and access to healthcare

Socioeconomic status is a major determinant of human health. Wealthier people generally enjoy better health and live longer than poorer people. Expenditure on healthcare services in developed countries dwarfs that in developing countries (Figure 17.4). At the global scale, a random sample of 20 countries shows a close correlation between wealth and life expectancy (Figure 17.5). In Mozambique and Chad, two of the world's poorest countries, average life expectancy at birth is 41 and 48 years respectively. Exceptionally low life expectancy linked to the HIV/AIDS epidemic is also found in the world's poorest countries in southern Africa. In contrast, the world's richest countries in Figure 17.5 — Luxembourg, Greece and New Zealand — have average life expectancies around 80 years.

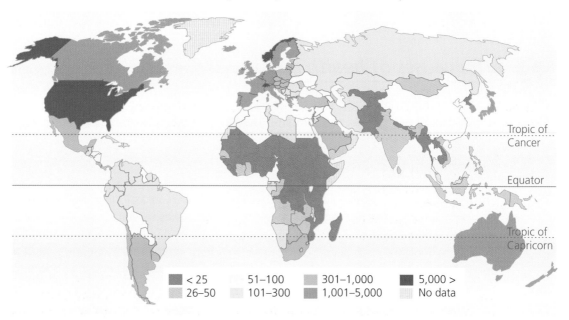

■ < 25		51–100	▨ 301–1,000	■ 5,000 >	
▨ 26–50		101–300	■ 1,001–5,000	▨ No data	

Figure 17.4 Healthcare spending, 2004 (per capita, US$)

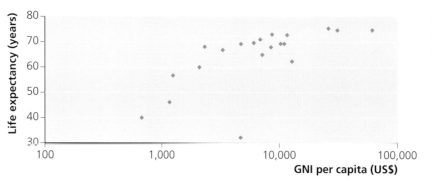

Figure 17.5 GNI per capita and life expectancy, 2009

CASE STUDY 1 The economic impact of HIV/AIDS

Thirty-three million people worldwide were infected with the HIV virus in 2007. In the same year, AIDS-related illnesses caused 2 million deaths. Since 1981, 25 million people have died from AIDS-related illnesses, most of them in Africa (Figure 17.6). The prevalence of HIV/AIDS has reduced rates of growth of per capita income by 0.7% a year and has had a detrimental effect on economic growth. The HIV/AIDS epidemic has greatly increased the total costs of treating and caring for patients, and has overburdened health services. In Ivory Coast, Zambia and Zimbabwe, HIV/AIDS patients occupy between 50% and 80% of all hospital beds. Countries like Botswana, which have been particularly hard hit by HIV/AIDS, suffer economically through reduced levels of saving, less investment and less productive employment.

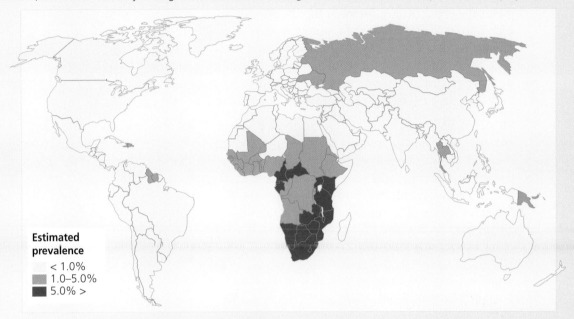

Estimated prevalence
- < 1.0%
- 1.0–5.0%
- 5.0% >

Figure 17.6 The global spread of HIV/AIDs, 2007

HIV/AIDS affects the most economically active and most productive members of society, reducing not only the size but also the quality of the workforce. The loss of skilled workers in health services and teaching has particularly severe economic effects. On an individual household level, HIV/AIDS:

- results in a loss of income
- forces children to leave school to care for dying parents
- increases household expenditure on medicines
- causes indebtedness
- leaves millions of children orphaned (11.6 million in Africa alone)
- greatly increases the likelihood of poverty

The relationship between socioeconomic status and health is also evident at regional and local scales. The most prosperous regions in the UK, the southeast and southwest, have average life expectancies of 78.9 years for males and 82.9 years for females. In Scotland, the least prosperous region in the UK, average life expectancies are 5 years lower. At a local scale, Kensington in central London, the wealthiest ward in the UK, has an average male life expectancy of 83.7 years. This compares with just 70.8 years in the poorest ward: Glasgow City.

Similar contrasts occur within large cities. In Bradford, Rombalds in the outer suburbs is the least deprived ward in the city and has the highest average life expectancy (80.7 years). Meanwhile, the most deprived ward, Bradford Moor in the inner city, also has the lowest life expectancy in Bradford (72.5 years).

It is clear that huge spatial variations in human health exist (as measured by life expectancy) at global, regional and local scales and that these differences are strongly influenced by the social and economic status of populations.

2.3 Geographical distance and access to healthcare services

At a regional scale, where people live often has a huge influence on their access to primary and secondary healthcare services. The difference between rural and urban communities is striking. Rural communities, which make up 19% of the UK's population, generally have poorer access to healthcare services. GP practices and community health teams in rural areas are likely to be located some distance from home and patients will have limited choice. Although nationally only 5% of patients in the UK have to travel more than 8 km to visit their GP, the proportion in rural areas is 12% (compared to 1% in urban areas). In rural northwest England, 88% of parishes have no GP surgery (Figure 17.7) and in Cumbria 41% live more than 12 km from a hospital. Moreover, journey times to healthcare services are often lengthened by poor roads.

This inequality of access hits some rural dwellers, such as older people, mothers with young children, disabled groups, those without a car and the poor, particularly hard. Because these groups cannot easily access health services, they make less use of them. The impact of such inequality means that for rural patients cancer may be diagnosed at a later stage, intervention rates for coronary heart disease are lower and rural patients are admitted to hospital less frequently than urban patients. The suggestion is that geography, by limiting access to healthcare services, is as much a determinant of ill health as genetic, environmental and lifestyle factors.

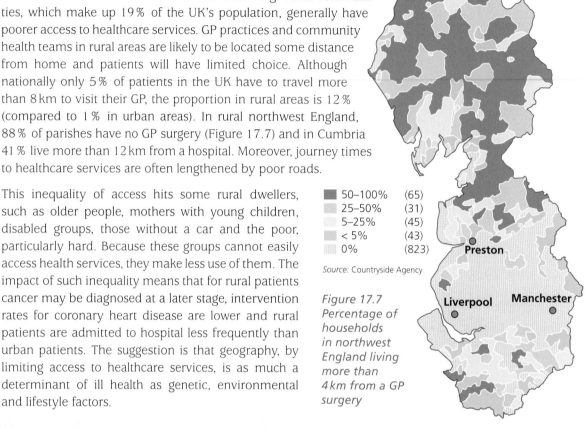

■ 50–100%	(65)
25–50%	(31)
5–25%	(45)
< 5%	(43)
0%	(823)

Source: Countryside Agency

Figure 17.7 Percentage of households in northwest England living more than 4 km from a GP surgery

Box 1 *Diffusion models*

Diffusion models simulate the geographical spread of infectious diseases such as influenza. They allow us to make predictions about the future spread of the disease over space. A flu epidemic begins when a handful of people are infected by a new strain of the flu virus. Initially, it spreads slowly, then it accelerates rapidly and, finally, when all susceptible people have been infected, it subsides. Over time the rate of transmission follows an S-shaped curve. Diffusion models assume that the probability of infection is inversely proportional to the distance between a carrier and those susceptible to the disease. As the disease spreads out from its origin, it occupies an ever larger geographical space. This is called **expansion diffusion**. The disease also moves through space as the population in the initial areas of infection either recover or succumb, with the result that the disease area relocates. Therfore, the spread of a disease is a combination of expansion and **relocation diffusion**.

The rate of diffusion depends on several factors:
■ transport technology and the speed of passenger travel
■ the density of the population (infectious diseases spread rapidly in overcrowded conditions)
■ the nature of the disease-carrying vector
■ wind direction (for diseases that are wind-borne)

In 2008, avian flu appeared in Europe, carried by migrating birds from Asia that can fly hundreds of kilometres in a few hours. Today, many diseases are spread rapidly around the globe by air transport. This is illustrated by the outbreak of swine flu in 2009. Swine flu was first detected in Mexico in April. Within a few days it had spread to the USA, by June people in 75 countries had been infected and, on 12 June, the World Health Organization declared it a pandemic. Its spread is a type of relocation (rather than expansion) diffusion, which often cascades down the settlement hierarchy from large cities with major international airports.

3 *Pollution and health risks*

3.1 Air pollution and health risks

Clean air is a basic requirement of human health. However, air pollution poses a significant threat to health worldwide. The WHO estimates that every year 2 million premature deaths occur through air pollution and millions more people suffer chronic ill health. Air pollution is primarily caused by burning fossil fuels, which emit pollutants like sulphur dioxide, nitrogen oxide and particulates. The main health problems are respiratory diseases such as bronchitis, asthma and lung cancer. More than half of the disease burden caused by air pollution is borne by populations in poor countries.

CASE STUDY 2	**Air pollution in northern China**
Background	China's principal cities and industrial regions suffer severe air pollution. This is due to China's rapid industrialisation in the past 30 years and its dependence on coal, the most polluting of fossil fuels, for electricity generation. The main pollutants are sulphur dioxide and particulates.
Shanxi province	The air pollution problem is critical in the heavy industrial region of Shanxi. Pollution comes from the huge concentration of heavy industries, especially coal mines, coke works, power stations, iron and steel, tar and chemical plants. The problems of air pollution are exacerbated by frequent dust storms and sand storms, which blow from the desertified interior. Of China's ten most polluted cities, four are in Shanxi. Notorious blackspots are the cities of Datong and Linfen. Coal burning is also responsible for acid rain, which falls on 30% of the country.
Impact	Linfen (pop. 3.5 million) is the most polluted city in China. Beilu on the city's outskirts is known locally as 'cancer village'. It is estimated that the death rate in this area is ten times higher than the average for China. Beilu is situated in a natural basin where air pollution from surrounding coke works, power stations and a pharmaceutical factory is trapped by temperature inversions. Other diseases related to airborne pollution are bronchitis and pneumonia and rates of lead poisoning are high. Levels of sulphur dioxide and particulates in cities in Shanxi exceed the WHO's standards many times. In Linfen, death rates are exceptionally high: for 55-year-olds and over, the rate has risen to 61 in 1,000.

3.2 Water pollution and health risks

The most common sources of water pollution that affect human health are domestic sewage and chemical effluent released by industry. Diseases like schistomiasis, related to inadequate sanitation and contaminated drinking water, have part of their life cycle in water. Some 160 million people are affected by the disease, which causes tens of thousands of deaths every year, mainly in sub-Saharan Africa. Most diarrhoeal diseases (including cholera) are also caused by poor sanitation and unsafe water supplies. Around 1.8 million people die every year from diarrhoeal diseases, of whom 90% are children under 5 years and most are in developing countries.

Health problems also result from the industrial pollution of water supplies. Inadequate controls on industrial emissions in poorer countries mean that river and groundwater pollution is often a major problem. In Bangladesh, between 28 million and 35 million people consume drinking water containing high levels of arsenic. In the worst cases this results in arsenicosis, a disease that causes skin lesions and carries with it a high risk of cancer. Fluorosis is common in heavily industrialised regions and is associated with coal-burning and smokestack industries. Excessive intake of fluorides damages bones and teeth. Fluorides transmitted to the environment through liquid effluent often accumulate to dangerous levels in aquatic food chains. Twenty-six million people in China suffer from dental fluorosis and there are over 1 million cases of skeletal fluorosis in the country, which is thought to be caused by polluted drinking water.

The Noyyal River in Tamil Nadu in southern India typifies the water pollution problem. Untreated liquid industrial wastes containing toxic chemicals and metals are discharged directly into the river from hundreds of dyeing and bleaching factories. Around 100 villages rely on the Noyyal for water supplies. One-quarter of their residents suffer from skin allergies, respiratory infections, gastritis and ulcers linked directly to river pollution.

3.3 Accidental pollution and health risks

The examples of pollution considered so far result from routine discharges of pollutants into the atmosphere and water bodies. Pollution is also caused by accidents. These one-off events can result in high fatality levels and chronic ill health. In 1984 a major pollution incident occurred at the Union Carbide pesticide factory at Bhopal, in India. A leak of 42 tonnes of deadly methyl isocyanate gas killed between 8,000 and 10,000 people within 72 hours. However, it is estimated that since the disaster a further 25,000 people may have died, with many more suffering injury and permanent disability.

CASE STUDY 3	The Chernobyl nuclear accident
Background	The world's worst nuclear accident occurred on 26 April 1986 at the Chernobyl nuclear power plant in Ukraine. The accident involved a meltdown of the reactor's core and a massive release of radiation.
Pollution hazard	The accident released radiation 100 times greater than the atomic bombs dropped on Hiroshima and Nagasaki. The radio-nuclides were carried by winds across northern Europe. Radioactive fallout affected not only Ukraine, Belarus and Russia, but also Scandinavia and northern Britain. Pollutants include uranium, plutonium, radioactive iodine, cesium-137 and strontium. Some radioactive pollutants have half-lives of thousands of years; others (e.g. cesium-137) are stored in soils, absorbed by vegetation and enter the food chain.
Impact	Millions of people across Europe were exposed to ionising radiation. Thyroid cancer in children living in the region around Chernobyl was (and remains) a major problem. Over 4,000 thyroid cases have been diagnosed since 2002. Most are linked to radioactive iodine found in milk. Between 1992 and 2002 a further 4,000 cases of thyroid cancer were detected in Belarus, Russia and Ukraine, mainly among those who were children at the time of the accident. Some experts believe the cancer rate has peaked. Others argue that it could take decades for all cancers to be detected.

3.4 Economic development, pollution and health risks

It is theorised that with economic development, pollution and associated health risks will undergo significant changes. In pre-industrial societies, environmental pollution and health risks are small. With industrial development, rapid expansion of the economy and population growth, pollution levels rise and mortality attributable to the effects of environment degradation increases. In post-industrial economies, with their emphasis on service activities, rising wealth and greater concern for the environment, pollution levels fall (Figure 17.9). These ideas are formalised in Kuznets environmental model (Box 2).

Box 2 *Kuznets environmental curve*

The Kuznets environmental curve (Figure 17.8) summarises the relationship between various indicators of environmental degradation (e.g. air and water pollution) and income per capita. In the early stages of economic growth, pollution increases because society gives little priority to environmental concerns. However, with growing wealth the trend reverses: greater value is placed on a clean environment. Action is taken to reduce and reverse the legacy of environmental pollution caused by industrialisation and further economic development leads to environmental improvement. The Kuznets model suggests that environmental pollution, and therefore human health, follow an inverted U shaped curve.

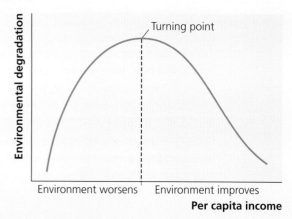

Figure 17.8 Kuznets curve

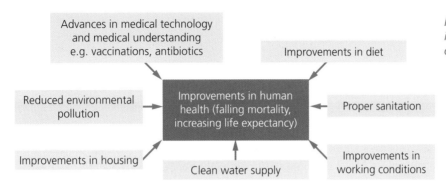

Figure 17.9 Progress in human health in developed countries

Advances in medical technology and medical understanding e.g. vaccinations, antibiotics

Improvements in diet

Reduced environmental pollution

Improvements in human health (falling mortality, increasing life expectancy)

Proper sanitation

Improvements in housing

Clean water supply

Improvements in working conditions

4 *Managing health risks*

4.1 The impact of ill health on people and communities

Malaria has a major economic impact on national economies and households in the developing world. The cost of prevention and treatment consumes scarce household resources. For example, spending on mosquito coils, aerosol sprays, bed nets and mosquito repellents in sub-Saharan Africa costs a family of five an average of US$55 a year. This is equal to 7% of a family's average annual income. In Nigeria, families affected by malaria spend 18% of their income on treatment. The burden of malaria also includes time spent caring for the ill and the reduction in well-being from sickness. For small farmers in Kenya, the economic impact of malaria amounts to a loss of income of between 9 and 18%. Malaria also places a huge burden on the public health sector. Because the disease takes a particularly heavy toll on children, it reduces the effectiveness of the workforce in future years and therefore harms economic development. Malaria costs Africa US$12 billion in lost GDP every year.

Obesity is a major health problem in developed countries and is linked to cancer, heart and liver disease and diabetes. In England, almost 1 in 4 adults are obese. Without intervention, by the mid-twenty-first century 9 out of 10 adults could be overweight or obese. The current cost of the obesity crisis to the NHS is £4.2 billion, a figure that could more than double by 2050. The cost to the wider economy in terms of loss of production was estimated to be £16 billion in 2008.

4.2 Management strategies and policies

Some health risks are harder to manage than others. Problems such as obesity and mental illness in developed countries reflect lifestyles and values in a modern consumer society. Large numbers of people lead sedentary lives. They prefer the convenience of fast food to home-cooked meals, they consume excessive amounts of alcohol and foods rich in saturated fat and they opt to travel by car rather than walking or cycling. Unhealthy lifestyles are often reinforced in the advertising of food by manufacturers and retailers.

In the context of societies based on consumerism, persuading millions of people to consume less, eat more healthily and take exercise is difficult. However, the UK government is making a coordinated effort to tackle the problems. In 2008 it announced plans to invest £372 million in 'healthy towns', where residents will be encouraged to cycle rather than drive. The plan also includes a £75 million advertising campaign to inform parents how they can improve their children's diets and exercise levels. Other initiatives will promote breastfeeding, review junk food advertising and restrict the amount of time children spend watching television and playing computer games. Ultimately, however, lifestyles that result in obesity are ones of choice and self-inflicted, which is why they are so difficult to tackle.

In contrast, people in developing countries more often suffer poor health from infectious diseases like malaria and are desperate for treatment, although obesity is becoming a problem in some developing countries. In a sense their health problems are easier to address. Even so, the sheer number of people suffering from diseases like malaria is daunting. The WHO reported 327 million cases of the disease in 2006, with nearly 1 million fatalities. International agencies like the WHO and World Bank fund expensive programmes to fight malaria in poor countries. In Senegal, volunteer workers operating under the

WHO's Global Malaria Programme distribute insect-treated nets and educate villagers about preventing and treating malaria. In Cambodia, malaria workers go from house to house, providing drugs and diagnosing and treating malaria. In 2008 the World Bank announced a US$1.1 billion expansion of its programme to eradicate malaria in Africa. The programme provides bed nets, drugs, insecticides and community education. Already there are signs of progress, with decreasing numbers of malaria cases and deaths in countries such as Eritrea, Ethiopia, Rwanda and Zambia.

USAID delivers the American government's official aid programme and is committed to improving global health. Its Global Health Bureau, with an annual budget of US$4.15 billion in 2007, supports health programmes in the developing world dedicated to fighting diseases such as HIV/AIDS, malaria and tuberculosis (TB). At a smaller scale, WaterAid is an international charity and non-governmental organisation (NGO) dedicated to overcoming poverty by helping the world's poorest people to access safe water, sanitation and hygiene education. It specialises in low-cost, sustainable solutions that can eradicate water-borne diseases and parasites such as diarrhoea and hookworms and improve human health and well-being.

4.3 Planning the health risk

Emergency disease relief

In the aftermath of major natural disasters, secondary hazards often develop such as disease epidemics. Because disaster planning is inadequate or non-existent in many developing countries, attempts to deal with epidemics following an earthquake, hurricane or flood are often uncoordinated and ineffective.

In October 2005 a major earthquake measuring 7.6 on the Richter scale devastated Kashmir in northern Pakistan and northern India. Some 87,000 people were killed and thousands of survivors were left without food, clean water and shelter. Pakistan was unprepared to deal with a disaster of this magnitude. Few disaster prevention measures were in place and the government was slow to respond and provide emergency aid. International emergency aid was provided by agencies such as the WHO and charities such as Oxfam and CARE.

Poor sanitation in overcrowded relief camps, which polluted surface streams, caused outbreaks of acute diarrhoeal disease, gastroenteritis and leishmaniasis (a parasitic skin disease). In addition, thousands of people traumatised by the quake suffered serious mental disorders including depression, psychosis and anxiety.

Planning to fight the swine flu pandemic

The response of governments to the swine flu outbreak in 2009 was carefully planned and coordinated. The WHO officially upgraded the flu outbreak, which began in Mexico in April 2009, as a pandemic in June 2009. However, the scale of the potential threat to human health had been apparent some weeks before the announcement and many governments already had plans in place to fight the spread of the disease.

In the UK the Department of Health issued guidelines and made plans to tackle the disease and protect the public. These included:

- informing the public about the symptoms of the disease and the precautions and actions needed to minimise its spread (e.g. staying at home, checking symptoms with the National Pandemic Flu Service)
- the need for testing to confirm the disease and then to trace all immediate contacts
- the closure of schools and other organisations where outbreaks were confirmed
- treating carriers and potential carriers of the disease with antiviral drugs (e.g. Tamiflu)
- stockpiling large supplies of antiviral drugs, sufficient to treat half the UK population in the worst-case scenario
- having sufficient supplies of antibiotics to treat flu cases where complications arise (e.g. secondary infections of the respiratory tract)
- sourcing sufficient vaccine to immunise those most at risk before the onset of the flu season (i.e. winter)